My Book

AF887787

This book belongs to

Name: ______________________________

Cover Design by :
Gowri Vemuri

First Edition :
May, 2020

Author :
Gowri Vemuri

Editor :
Ritvik Pothapragada

Questions: mathknots.help@gmail.com

This book is dedicated to:

My Mom, who is my best critic, guide and supporter.

To what I am today, and what I am going to become tomorrow,

is all because of your blessings, unconditional affection and support.

This book is dedicated to the

strongest women of my life ,

my dearest mom

and

to all those moms in this universe.

G.V.

The percentage symbol is a representation of percentage. In statistics, percentages are often left in their base form of 0 - 1, where 1 represents the whole. We multiply the decimal by a factor of 100 to find the percentage.

Percent Equation form :

Percentages can be setup as proportions. The parts of a percent can be found by setting up a proportion as below.

$$\frac{\text{is}}{\text{of}} = \frac{\text{percent}}{100} \quad \text{OR} \quad \frac{\text{part}}{\text{total}} = \frac{\text{percent}}{100}$$

Before we calculate a percentage, we should understand exactly what a percentage is ? The word percentage comes from the word percent. If you split the word percent into its root words, you see "per" and "cent." Cent is an old European word with French, Latin, and Italian origins meaning "hundred". So, percent is translated directly to "per hundred." If we have 39 percent, we literally have 39 per 100. If it snowed 16 times in the last 100 days, it snowed 16 percent of the time.

Whole numbers, decimals and fractions are converted to percentages. Decimal format is easier to convert into a percentage. Fractions can be converted into decimals and then to percentages.

Converting a decimal to a percentage is as simple as multiplying it by 100. To convert 0.92 to a percent, simply multiple 0.92 by 100.

$0.92 \times 100 = 92\%$

A percentage is an expression of part of the whole. Nothing is represented by 0%, and the whole amount is 100%. Everything else is somewhere in between 0 and 100.

For example, say you have 20 muffins. If you share 12 muffins with your friends, then you have shared 12 out of the 20 muffins ($\frac{12}{20} \times 100\% = 60\%$ shared). If 20 muffins are 100% and you share 60% of them, then 100% - 60% = 40% of the muffins are still left.

Determine the part of the whole :

Given the value for part of the whole and the whole.
OR
Given two parts that make up the whole.

It is important to differentiate what the percentage is "of."
Example : A jar contains 99 gold beads and 51 blue beads. A total of 150 beads are in the jar. In this example, 150 beads makes up a whole jar of beads, i.e. 100%.

1. **Put the two values into a fraction.** The part goes on top of the fraction (numerator), and the whole goes on the bottom (denominator). Therefore the fraction in this case is $\frac{99}{150}$ (part/whole) or $\frac{51}{150}$ (part/whole).

2. **Convert the fraction into a decimal.** Convert $\frac{99}{150}$ into decimal, divide 99 by 150

$$\frac{99}{150} = 0.66$$

Converting the decimal into a percent :

Multiply the result obtained in the step above by 100% (per 100 = *per cent*).
0.66 multiplied by 100 equals 66%. Gold beads are 66% of the total beads in the jar.

Converting the percentage into a decimal : Working backward from before, divide the percentage by 100, or you can multiply by 0.01.

$$66\% = \frac{66}{100} = 0.66$$

Re-word the problem with your new values :

Given in the form of "**X** of **Y** is **Z**." X is the decimal form of your percent, "of" means to multiply, Y is the whole amount, and Z is the answer.

Example: 3% of $25 is 0.75.
The amount of interest accrued each day on a 3% of loan amount $25 s 0.75

Discounts :

Discounts are offered on the original price of the item on sale.

1. The discount percent need to be subtracted from the whole, meaning if a discount of 20% is offered on an item priced at \$60 then sale price is 80% of \$60.
   100% - 20% = 80% (whole percent - discount percent).

   $80\% \text{ of } \$60 = \frac{80}{100} \text{ X } 60 = \$48$

Markup percentage :

Markup percentage is also referred as gain percentage. Markups are added to the original price of the item before selling it. The greater the markup percentage higher are the profits.

1. The markup percent needs to be added to the whole, meaning if a markup of 20% is added on an item priced at \$60 then sale price is 120% of \$60.
   100% + 20% = 120% (whole percent + markup percent).

   $120\% \text{ of } \$60 = \frac{120}{100} \text{ X } 60 = \$72$

Tax percentage :

Taxes are added to the original price of the item before selling it to the customer.

1. The tax percent needs to be added to the whole, meaning if a tax of 5% is added on an item priced at \$60 then sale price is 105% of \$60.
   100% + 5% = 105% (whole percent + tax percent).

   $105\% \text{ of } \$60 = \frac{105}{100} \text{ X } 60 = \$63$

Simple Interest :

Simple interest is an easy way of calculating interest on a loan amount.

I = PTR

I = Simple Interest
P = Principle amount of loan
T = Term of the loan in years
R = Rate of interest in decimal form.

Remember final loan amount (A) = P + I

Or

$$I = \frac{PTR}{100}$$

I = Simple Interest
P = Principle amount of loan
T = Term of the loan in years
R = Rate of interest in numerical value only (not in decimal form since the formula has 100 in the denominator).

Remember final loan amount (A) = P + I

Example : Dan took a loan of $1,650 at 7% rate of interest for 5 years from his friend Johnson. Using simple interest find the loan amount to be paid to Johnson at the end of the term.

$$I = \frac{PTR}{100}$$

P = $1,650
R = 7
T = 5 years

$$I = \frac{1650 \text{ X } 7 \text{ X } 5}{100}$$

$$I = \frac{57750}{100}$$

$$= \$577.50$$

Interest to be paid at the end of the loan term is $577.50
Remember the original loan amount $1,650 also need to be paid of.
Amount to be paid of (A) = P + I

$$A = \$1,650 + \$577.50$$
$$= \$2,227.50$$

Compound interest :

Compound interest is a complex way of calculating interest on a loan amount and the interest on the interest accurred.

$$A = P(1 + \frac{r}{n})^{nt}$$

Where A = Amount at the end of the loan term
Note :It includes principle amount and the interest

P = Principle loan amount
r = Rate of loan or loan percent or interest rate.
Note : Rate needs to be converted to decimal.

n = Number of times interest is compounded per year.
t = Loan term in years.

Remember final loan amount (A) = P + I

OR

$$A = P(1 + \frac{r}{100n})^{nt}$$

Where A = Amount at the end of the loan term
Note :It includes principle amount and the interest

P = Principle loan amount
r = Rate of loan or loan percent or interest rate.
Note : Rate need not be converted to decimal.

n = Number of times interest is compounded per year.
t = Loan term in years.

Remember final loan amount (A) = P + I

Example : Dan took a loan of \$49,100 at 14% rate of interest for 2 years compounded annually from his friend Johnson. Using compound interest find the loan amount to be paid to Johnson at the end of the term.

$$A = P\left(1 + \frac{r}{100n}\right)^{nt}$$

P = \$49,100
R = 14
T = 2 years
n = 1

$$A = 49100\left(1 + \frac{14}{100}\right)^{2}$$

$$A = 49100\,(1.14)^{2}$$

$$A = \$63,810.36$$

$$A = P + I \text{ OR } I = A - P$$

Interest I = \$63,810.36 - \$49,100 = \$14,710.36

Interest to be paid at the end of the loan term is \$14,710.36
Remember the original loan amount \$49,100 also need to be paid of.
Amount to be paid of (A) = P + I

$$A = \$63,810.36$$

Dan needs to pay \$63,810.36 to Johnson at the end of the loan term.

Example : Dan took a loan of $49,100 at 14% rate of interest for 2 years compounded semi annually from his friend Johnson. Using compound interest find the loan amount to be paid to Johnson at the end of the term.

$$A = P\left(1 + \frac{r}{100n}\right)^{nt}$$

P = $49,100
R = 14
T = 2 years
n = 2

$$A = 49100\left(1 + \frac{14}{200}\right)^{4}$$

$$A = 49100\,(1.07)^{4}$$

A = $64,360.08

A = P + I OR I = A - P

Interest I = $64,360.08 - $49,100 = $15,260.08

Interest to be paid at the end of the loan term is $15,260.08
Remember the original loan amount $49,100 also need to be paid of.
Amount to be paid of (A) = P + I

A = $64,360.08

Dan needs to pay $64,360.08 to Johnson at the end of the loan term.

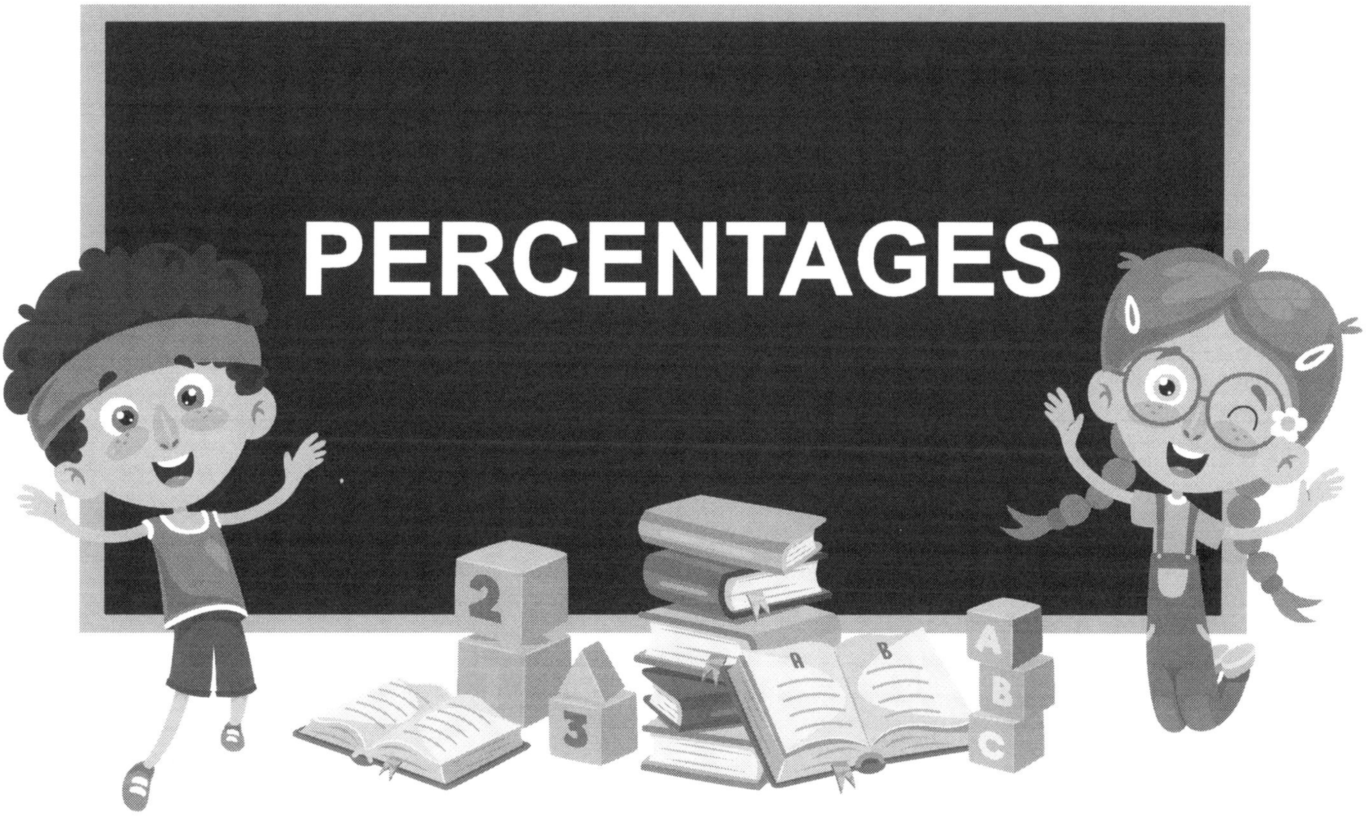
PERCENTAGES

Convert the below fractions to decimals.

(1) $\frac{33}{125}$

(2) $3\frac{1}{8}$

(3) $\frac{323}{500}$

(4) $\frac{47}{100}$

(5) $\frac{11}{20}$

(6) $\frac{1}{3}$

(7) $\frac{3}{125}$

(8) $\frac{47}{1000}$

(9) $\frac{89}{100}$

(10) $\frac{8}{9}$

Convert the below fractions to decimals.

(11) $\frac{63}{100}$

(12) $\frac{5}{16}$

(13) $\frac{21}{25}$

(14) $\frac{74}{125}$

(15) $\frac{3}{5}$

(16) $\frac{9}{11}$

(17) $\frac{7}{1000}$

(18) $\frac{893}{1000}$

(19) $\frac{11}{25}$

(20) $\frac{1}{100}$

Convert the below fractions to decimals.

(21) $\frac{1}{25}$

(22) $\frac{1}{4}$

(23) $7\frac{7}{100}$

(24) $\frac{2}{3}$

(25) $7\frac{21}{100}$

(26) $7\frac{1}{4}$

(27) $\frac{1}{2}$

(28) $\frac{3}{4}$

(29) $\frac{2}{333}$

(30) $\frac{2}{5}$

Convert the below fractions to decimals.

(31) $\frac{9}{10}$

(32) $\frac{23}{33}$

(33) $\frac{17}{50}$

(34) $\frac{1}{200}$

(35) $\frac{39}{250}$

(36) $\frac{20}{33}$

(37) $\frac{4}{11}$

(38) $\frac{1}{125}$

(39) $\frac{1}{5}$

(40) $\frac{843}{1000}$

Convert the below fractions to decimals.

(41) $\frac{7}{10}$

(42) $2\frac{2}{3}$

(43) $9\frac{99}{100}$

(44) $\frac{7}{9}$

(45) $\frac{4}{5}$

(46) $\frac{38}{99}$

(47) $\frac{7}{33}$

(48) $\frac{7}{8}$

(49) $\frac{5}{8}$

(50) $\frac{3}{10}$

Convert the below decimals to fractions.

(51) 0.81

(52) 0.35

(53) 0.722

(54) 0.828

(55) 0.008

(56) $0.\overline{0060}$

(57) 0.7

(58) 0.875

(59) $0.\overline{43}$

(60) $1.\overline{3}$

Convert the below decimals to fractions.

(61) 0.2

(62) 0.375

(63) $0.\overline{4587}$

(64) 0.341

(65) $0.\overline{003}$

(66) 0.24

(67) 0.8

(68) 0.125

(69) $0.\overline{09}$

(70) 6.9

Convert the below decimals to fractions.

(71) 0.5

(72) 3.5

(73) 0.76

(74) 0.21

(75) $0.\overline{1}$

(76) 0.0075

(77) 4.4

(78) 0.33

(79) $0.\overline{5}$

(80) 0.75

Convert the below decimals to fractions.

(81) 6.25

(82) 0.12

(83) $0.\overline{3}$

(84) 0.3

(85) $0.\overline{002}$

(86) 0.01

(87) 0.22

(88) 3.3

(89) $0.\overline{78}$

(90) 0.4

Convert the below decimals to fractions.

(91) $0.\overline{6}$

(92) $0.\overline{8}$

(93) 0.37

(94) 0.6

(95) 0.455

(96) $2.\overline{6}$

(97) 0.03

(98) 0.9

(99) 0.25

(100) 8.7

Convert the below fractions to percentages (Pay attention to repeating decimals).

(101) $\frac{11}{100}$

(102) $\frac{1}{1000}$

(103) $\frac{9}{10}$

(104) $\frac{8}{9}$

(105) $\frac{2}{999}$

(106) $\frac{1}{3}$

(107) $\frac{7}{10}$

(108) $\frac{1}{2}$

(109) $\frac{351}{500}$

(110) $6\frac{2}{3}$

Convert the below fractions to percentages (Pay attention to repeating decimals).

(111) $\frac{20}{33}$

(112) $2\frac{1}{8}$

(113) $\frac{19}{33}$

(114) $\frac{1}{100}$

(115) $\frac{76}{99}$

(116) $\frac{4}{11}$

(117) $\frac{2}{11}$

(118) $\frac{87}{100}$

(119) $\frac{3}{8}$

(120) $\frac{14}{33}$

Convert the below fractions to percentages (Pay attention to repeating decimals).

(121) $\frac{1}{5}$

(122) $\frac{46}{99}$

(123) $\frac{23}{33}$

(124) $1\frac{1}{2}$

(125) $\frac{1}{250}$

(126) $\frac{1}{125}$

(127) $\frac{73}{100}$

(128) $\frac{3}{10}$

(129) $2\frac{3}{4}$

(130) $\frac{109}{125}$

Convert the below fractions to percentages (Pay attention to repeating decimals).

(131) $\frac{7}{8}$

(132) $\frac{3}{5}$

(133) $\frac{88}{333}$

(134) $\frac{5}{8}$

(135) $\frac{25}{33}$

(136) $\frac{1}{10}$

(137) $\frac{4}{5}$

(138) $\frac{39}{100}$

(139) $\frac{2}{5}$

(140) $\frac{10}{11}$

Convert the below fractions to percentages (Pay attention to repeating decimals).

(141) $6\frac{1}{5}$

(142) $\frac{56}{99}$

(143) $\frac{3}{4}$

(144) $\frac{5}{11}$

(145) $\frac{3}{40}$

(146) $3\frac{3}{5}$

(147) $\frac{373}{500}$

(148) $\frac{1}{8}$

(149) $\frac{1}{4}$

(150) $\frac{1}{333}$

Convert the below fractions to percentages (Pay attention to repeating decimals).

(151) $\frac{8}{99}$

(152) $\frac{1}{200}$

(153) $\frac{3}{11}$

(154) $5\frac{7}{8}$

(155) $7\frac{1}{4}$

(156) $\frac{173}{200}$

(157) $\frac{43}{100}$

(158) $\frac{1}{33}$

(159) $\frac{6}{11}$

(160) $\frac{2}{3}$

Convert the below fractions to percentages (Pay attention to repeating decimals).

(161) $1\frac{9}{10}$

(162) $3\frac{1}{4}$

(163) $\frac{15}{16}$

(164) $\frac{17}{35}$

(165) $6\frac{2}{5}$

(166) $\frac{1}{100}$

(167) $\frac{11}{20}$

(168) $2\frac{1}{2}$

(169) $\frac{27}{50}$

(170) $3\frac{1}{8}$

Convert the below fractions to percentages (Pay attention to repeating decimals).

(171) $\frac{3}{50}$

(172) $\frac{5}{8}$

(173) $\frac{29}{30}$

(174) $\frac{63}{100}$

(175) $\frac{9}{17}$

(176) $6\frac{1}{10}$

(177) $\frac{69}{70}$

(178) $\frac{1}{300}$

(179) $\frac{2}{5}$

(180) $\frac{1}{2}$

Convert the below fractions to percentages (Pay attention to repeating decimals).

(181) $\frac{1}{199}$

(182) $\frac{1}{106}$

(183) $\frac{7}{17}$

(184) $\frac{9}{25}$

(185) $\frac{1}{9}$

(186) $\frac{41}{75}$

(187) $1\frac{1}{2}$

(188) $\frac{1}{60}$

(189) $\frac{3}{4}$

(190) $\frac{71}{100}$

Convert the below fractions to percentages (Pay attention to repeating decimals).

(191) $\frac{1}{270}$

(192) $\frac{37}{65}$

(193) $\frac{3}{500}$

(194) $\frac{3}{8}$

(195) $5\frac{1}{2}$

(196) $\frac{1}{163}$

(197) $\frac{1}{160}$

(198) $\frac{1}{115}$

(199) $2\frac{3}{4}$

(200) $\frac{33}{100}$

Convert the below fractions to percentages (Pay attention to repeating decimals).

(201) $\frac{1}{207}$

(202) $9\frac{2}{5}$

(203) $\frac{1}{35}$

(204) $\frac{3}{25}$

(205) $\frac{19}{20}$

(206) $\frac{1}{4}$

(207) $\frac{7}{30}$

(208) $\frac{1}{900}$

(209) $\frac{7}{10}$

(210) $\frac{1}{10}$

Convert the below fractions to percentages (Pay attention to repeating decimals).

(211) $\frac{3}{380}$

(212) $\frac{1}{5}$

(213) $\frac{58}{75}$

(214) $\frac{77}{100}$

(215) $8\frac{5}{8}$

(216) $\frac{1}{8}$

(217) $\frac{3}{5}$

(218) $\frac{18}{19}$

(219) $\frac{81}{100}$

(220) $\frac{4}{5}$

Convert the below decimals to percentages.

(221) 0.96

(222) 0.0066

(223) 0.0005

(224) 0.35

(225) 0.9

(226) 0.14

(227) 0.5

(228) 0.0069

(229) 0.0058

(230) 0.39

Convert the below decimals to percentages.

(231) 0.383

(232) 0.84

(233) 0.01

(234) 0.47

(235) 0.358

(236) 0.08

(237) 0.02

(238) 0.51

(239) 0.343

(240) 0.0094

Convert the below decimals to percentages.

(241) 0.0089

(242) 0.394

(243) 0.2

(244) 0.218

(245) 0.487

(246) 0.29

(247) 0.62

(248) 0.27

(249) 0.06

(250) 0.7

Convert the below decimals to percentages.

(251) 0.0037

(252) 0.00903

(253) 0.1

(254) 0.67

(255) 0.8

(256) 0.453

(257) 0.07

(258) 0.993

(259) 0.0088

(260) 0.932

Convert the below decimals to percentages.

(261) 0.6

(262) 0.09

(263) 0.52

(264) 0.4

(265) 0.38

(266) 0.04

(267) 0.0023

(268) 0.75

(269) 0.167

(270) 0.347

Convert the below decimals to percentages.

(271) 0.05

(272) 0.88

(273) 0.00305

(274) 0.518

(275) 0.77

(276) 0.21

(277) 0.81

(278) 0.18

(279) 0.63

(280) 0.021

Convert the below decimals to percentages.

(281) 1.86

(282) 0.42

(283) 0.94

(284) 0.32

(285) 0.03

(286) 0.98

(287) 0.66

(288) 7.58

(289) 0.16

(290) 1.06

Convert the below decimals to percentages.

(291) 0.09

(292) 2.59

(293) 2.21

(294) 0.2

(295) 0.8

(296) 0.69

(297) 0.85

(298) 0.64

(299) 0.36

(300) 0.88

Convert the below decimals to percentages.

(301) 0.77

(302) 7.79

(303) 3.33

(304) 0.53

(305) 0.5

(306) 3.88

(307) 4.88

(308) 1.63

(309) 0.4

(310) 2.09

Convert the below decimals to percentages.

(311) 0.9

(312) 3.24

(313) 0.17

(314) 0.11

(315) 9.62

(316) 0.59

(317) 0.52

(318) 0.01

(319) 6.9

(320) 0.27

Convert the below decimals to percentages.

(321) 0.05

(322) 4.77

(323) 0.02

(324) 0.22

(325) 0.45

(326) 4.95

(327) 0.1

(328) 0.91

(329) 0.06

(330) 0.19

Convert the below decimals to percentages.

(331) 2.44

(332) 1.61

(333) 0.15

(334) 0.44

(335) 0.04

(336) 0.87

(337) 0.41

(338) 5.1

(339) 0.89

(340) 5.09

Find the percentage of the below.

(341) What percent of 61.7 ft is 50 ft?

(342) What percent of 39 ft is 37 ft?

(343) $127 is what percent of $64?

(344) 45.2 ft is what percent of 85 ft?

(345) What percent of 76 ft is 37 ft?

Find the percentage of the below.

(346) What percent of 113 m is 44 m?

(347) What percent of 45 ft is 113 ft?

(348) 87 grams is what percent of 116 grams?

(349) What percent of 129.3 minutes is 9.9 minutes?

(350) 123 hours is what percent of 100 hours?

Find the percentage of the below.

(351) What percent of 48 inches is 86 inches?

(352) What percent of 60 minutes is 105 minutes?

(353) What percent of $89 is $64?

(354) What percent of 25 tons is 156 tons?

(355) What percent of 136 m is 55.4 m?

Find the percentage of the below.

(356) 16 grams is what percent of 148 grams?

(357) What percent of 74 ft is 18 ft?

(358) 134 hours is what percent of 149.9 hours?

(359) What percent of 129 miles is 109.7 miles?

(360) 156 minutes is what percent of 78 minutes?

Find the percentage of the below.

(361) $74 is what percent of $82?

(362) What percent of 131 inches is 112 inches?

(363) 66 grams is what percent of 62 grams?

(364) What percent of 62 grams is 105 grams?

(365) 126 hours is what percent of 136 hours?

Find the percentage of the below.

(366) What percent of 101 tons is 27 tons?

(367) What percent of 98 m is 37 m?

(368) 60 m is what percent of 79 m? (

(369) What percent of 124.4 minutes is 87 minutes?

(370) 1 mile is what percent of 112.5 miles?

Find the percentage of the below.

(371) 86 minutes is what percent of 103.6 minutes?

(372) 4 inches is what percent of 42 inches?

(373) 72 miles is what percent of 157 miles?

(374) What percent of 115 hours is 100 hours?

(375) What percent of 154 inches is 112 inches?

Find the percentage of the below.

(376) 21 grams is what percent of 71 grams?

(377) 100.6 ft is what percent of 137 ft?

(378) $1 is what percent of $65?

(379) 42 hours is what percent of 103 hours?

(380) 77 grams is what percent of 140 grams?

Find the percentage of the below.

(381) 15 tons is what percent of 100 tons?

(382) 54 minutes is what percent of 110 minutes?

(383) 155.3 minutes is what percent of 160 minutes?

(384) 146 grams is what percent of 146 grams?

(385) $40 is what percent of $101?

Find the percentage of the below.

(386) What percent of 155 tons is 4.9 tons?

(387) 7 miles is what percent of 38 miles?

(388) What percent of 70 miles is 9.5 miles?

(389) 46 m is what percent of 120 m?

(390) 28 tons is what percent of 90 tons?

Find the percentage of the below.

(391) What percent of 149.5 m is 107 m?

(392) What percent of 157 ft is 2.4 ft?

(393) What percent of 77 miles is 23 miles?

(394) What percent of 124 ft is 29.5 ft?

(395) What percent of 81 ft is 69 ft?

Find the percentage of the below.

(396) 42 m is what percent of 105 m?

(397) 82 m is what percent of 21 m?

(398) What percent of 144 miles is 19 miles?

(399) 8 miles is what percent of 3.4 miles?

(400) 44 inches is what percent of 150.5 inches?

Find the percentage of the below.

(401) What percent of 150 miles is 159 miles?

(402) 23 ft is what percent of 43 ft?

(403) What percent of 91 m is 7.7 m?

(404) 9 m is what percent of 92 m?

(405) 46 tons is what percent of 53.6 tons?

Find the percentage of the below.

(406) What percent of 151 minutes is 5.3 minutes?

(407) 155 hours is what percent of 98 hours?

(408) 16.1 grams is what percent of 121.2 grams?

(409) What percent of 113.7 minutes is 64 minutes?

(410) What percent of 127.8 m is 82 m?

Find the percentage of the below.

(411) 1 ton is what percent of 74 tons?

(412) What percent of 151 m is 17 m?

(413) What percent of 63 minutes is 106 minutes?

(414) 34 inches is what percent of 100 inches?

(415) What percent of 23 miles is 132 miles?

Find the percentage of the below.

(416) What percent of $155 is $32?

(417) 63 tons is what percent of 128 tons?

(418) What percent of 100 minutes is 22 minutes?

(419) What percent of 68 hours is 14 hours?

(420) 27 grams is what percent of 108.8 grams?

Find the part or whole of the given percentage as below.

(421) What is 360% of 116 ft?

(422) 83% of 92 inches is what?

(423) 95% of 59 minutes is what?

(424) What is 40% of 22 hours?

(425) What is 180% of 120 hours?

Find the part or whole of the given percentage as below.

(426) 56% of 148 tons is what?

(427) 70% of 25 m is what?

(428) What is 95% of 42.5 minutes?

(429) What is 38% of 119 ft?

(430) 10% of 8 hours is what?

Find the part or whole of the given percentage as below.

(431) What is 76% of 35 tons?

(432) What is 62% of 73.7 grams?

(433) 79% of 25 tons is what?

(434) What is 76% of 37 grams?

(435) 71% of 152 inches is what?

Find the part or whole of the given percentage as below.

(436) What is 99% of 54 grams?

(437) 9% of 66 tons is what?

(438) What is 32% of $22?

(439) What is 32% of 148 grams?

(440) What is 166% of 80 inches?

Find the part or whole of the given percentage as below.

(441) 70% of 147 m is what?

(442) What is 10% of 22.3 hours?

(443) What is 171% of 94 m?

(444) 4% of $9 is what?

(445) What is 270% of 124 minutes?

Find the part or whole of the given percentage as below.

(446) What is 200% of 11 grams?

(447) 96% of $90 is what?

(448) 3% of 76.1 m is what?

(449) What is 84% of 131 ft?

(450) What is 78% of 50.8 tons?

Find the part or whole of the given percentage as below.

(451) 46% of 105 inches is what?

(452) What is 86% of $147?

(453) What is 94% of 143 ft?

(454) 3% of 91 tons is what?

(455) 96% of 159.6 miles is what?

Find the part or whole of the given percentage as below.

(456) 64% of 157 ft is what?

(457) 9% of 53 grams is what?

(458) 74% of $74 is what?

(459) 52% of 22.4 tons is what?

(460) 370% of 106.1 tons is what?

Find the part or whole of the given percentage as below.

(461) What is 26% of 36 minutes?

(462) What is 106% of 139 minutes?

(463) What is 86% of 117 grams?

(464) 170% of 106 tons is what?

(465) 32% of 93 inches is what?

Find the part or whole of the given percentage as below.

(466) 89% of 85 grams is what?

(467) 90% of 29.4 tons is what?

(468) What is 320% of 105 tons?

(469) 350% of 94 m is what?

(470) 3% of 130 minutes is what?

Find the part or whole of the given percentage as below.

(471) 240% of $66 is what?

(472) 64% of $12 is what?

(473) 5% of 86 m is what?

(474) 70% of 49 grams is what?

(475) What is 90% of 135 minutes?

Find the part or whole of the given percentage as below.

(476) 57% of 36 miles is what?

(477) 41% of 47 inches is what?

(478) 1% of 87 ft is what?

(479) 16% of 98 minutes is what?

(480) 24% of 127 m is what?

Find the part or whole of the given percentage as below.

(481) 79% of 98 minutes is what?

(482) 39% of 135.7 m is what?

(483) What is 18% of 8 inches?

(484) 56% of 139 minutes is what?

(485) 16% of 68 inches is what?

Find the part or whole of the given percentage as below.

(486) 38% of 8.5 grams is what?

(487) What is 80% of $123?

(488) What is 57% of 41 tons?

(489) What is 15% of $37?

(490) 380% of 111 m is what?

Find the part or whole of the given percentage as below.

(491) 82% of 153 miles is what?

(492) 330% of 4.8 inches is what?

(493) 320% of 87 ft is what?

(494) 81% of 38 m is what?

(495) What is 220% of 133 minutes?

Find the part or whole of the given percentage as below.

(496) 88% of $105 is what?

(497) What is 58% of 133 inches?

(498) What is 12% of 104.4 tons?

(499) 41% of 64 miles is what?

(500) 98% of 86.3 miles is what?

Find the part or whole of the given percentage as below.

(501) 56 is 40% of what?

(502) 53% of what is 150?

(503) 18 is 230% of what?

(504) 29% of what is 67?

(505) 250% of what is 156?

Find the part or whole of the given percentage as below.

(506) 29% of what is 32?

(507) 30 is 73% of what?

(508) 73% of what is 76.8?

(509) 134 is 95% of what?

(510) 59% of what is 102?

Find the part or whole of the given percentage as below.

(511) 25% of what is 59?

(512) 28% of what is 136?

(513) 152 is 78% of what?

(514) 47 is 5% of what?

(515) 71% of what is 107?

Find the part or whole of the given percentage as below.

(516) 56% of what is 107?

(517) 37 is 94% of what?

(518) 1% of what is 94?

(519) 4 is 27% of what?

(520) 400% of what is 29?

Find the part or whole of the given percentage as below.

(521) 34% of what is 147.1?

(522) 40 is 56% of what?

(523) 59% of what is 38?

(524) 82 is 54% of what?

(525) 44% of what is 91?

Find the part or whole of the given percentage as below.

(526) 20 is 76% of what?

(527) 27% of what is 36?

(528) 54 is 97% of what?

(529) 116 is 45% of what?

(530) 32 is 51% of what?

Find the part or whole of the given percentage as below.

(531) 35 is 36% of what?

(532) 79% of what is 14?

(533) 85 is 110% of what?

(534) 194% of what is 146.2?

(535) 102 is 120% of what?

Find the part or whole of the given percentage as below.

(536) 84 is 250% of what?

(537) 29% of what is 8.4?

(538) 6% of what is 112?

(539) 5% of what is 147?

(540) 49 is 3% of what?

Find the part or whole of the given percentage as below.

(541) 8 is 95% of what?

(542) 99 is 14% of what?

(543) 31.7 is 32% of what?

(544) 37 is 34% of what?

(545) 36% of what is 149?

Find the part or whole of the given percentage as below.

(546) 13% of what is 150?

(547) 56.1 is 69% of what?

(548) 41% of what is 58?

(549) 150 is 50% of what?

(550) 134 is 147% of what?

Find the part or whole of the given percentage as below.

(551) 99% of what is 68?

(552) 9 is 80% of what?

(553) 5% of what is 28?

(554) 94% of what is 73?

(555) 70% of what is 154?

Find the part or whole of the given percentage as below.

(556) 99% of what is 159?

(557) 159 is 57% of what?

(558) 94 is 89% of what?

(559) 42 is 390% of what?

(560) 37% of what is 58?

Find the part or whole of the given percentage as below.

(561) 6% of what is 34.3?

(562) 65 is 270% of what?

(563) 67% of what is 127?

(564) 9 is 25% of what?

(565) 47 is 95% of what?

Find the part or whole of the given percentage as below.

(566) 53% of what is 30?

(567) 86.8 is 99% of what?

(568) 94% of what is 114?

(569) 89.6 is 33% of what?

(570) 90 is 170% of what?

Find the part or whole of the given percentage as below.

(571) 230% of what is 105.7?

(572) 98% of what is 4.1?

(573) 350% of what is 43?

(574) 26% of what is 78?

(575) 53 is 2% of what?

Find the part or whole of the given percentage as below.

(576) 78% of what is 99?

(577) 49% of what is 4?

(578) 58.1 is 66% of what?

(579) 400% of what is 112?

(580) 103 is 340% of what?

Find the percent change of the below and state if it's a percentage increase or decrease. (Round to the nearest tenth)

(581) From 47 to 138

(582) From 45 to 42

(583) From 83 to 181

(584) From 44 to 174

(585) From 85 to 55

Find the percent change of the below and state if it's a percentage increase or decrease. (Round to the nearest tenth)

(586) From 40 to 165

(587) From 111 to 8

(588) From 131 to 174

(589) From 27 to 114

(590) From 139 to 21

Find the percent change of the below and state if it's a percentage increase or decrease. (Round to the nearest tenth)

(591) From 164 to 21

(592) From 24 to 55

(593) From 54 to 149

(594) From 174 to 127

(595) From 159 to 79

Find the percent change of the below and state if it's a percentage increase or decrease. (Round to the nearest tenth)

(596) From 86 to 56.6

(597) From 117 to 122

(598) From 67 to 7.3

(599) From 70 to 81

(600) From 107 to 167

Find the percent change of the below and state if it's a percentage increase or decrease. (Round to the nearest tenth)

(601) From 139 to 109

(602) From 167 to 50

(603) From 67 to 105

(604) From 13 to 44

(605) From 78 to 100

Find the percent change of the below and state if it's a percentage increase or decrease. (Round to the nearest tenth)

(606) From 49 to 93

(607) From 53 to 198

(608) From 57 to 59

(609) From 65.8 to 154

(610) From 185 to 151

Find the percent change of the below and state if it's a percentage increase or decrease. (Round to the nearest tenth)

(611) From 21 to 11

(612) From 180 to 169

(613) From 112 to 117

(614) From 51 to 4

(615) From 63 to 166

Find the percent change of the below and state if it's a percentage increase or decrease. (Round to the nearest tenth)

(616) From 142 to 51

(617) From 52 to 39

(618) From 133 to 199

(619) From 66 to 92

(620) From 141 to 129

Find the percent change of the below and state if it's a percentage increase or decrease. (Round to the nearest tenth)

(621) From 148 to 141

(622) From 40 to 151

(623) From 198 to 58

(624) From 9 to 28

(625) From 128 to 40

Find the percent change of the below and state if it's a percentage increase or decrease. (Round to the nearest tenth)

(626) From 83 to 89

(627) From 95 to 165

(628) From 138 to 57

(629) From 103 to 145

(630) From 197 to 82

Find the percent change of the below and state if it's a percentage increase or decrease. (Round to the nearest tenth)

(631) From 195 to 95

(632) From 105 to 12

(633) From 62 to 47

(634) From 181 to 136

(635) From 195 to 146

Find the percent change of the below and state if it's a percentage increase or decrease. (Round to the nearest tenth)

(636) From 133 to 185

(637) From 180 to 36

(638) From 113 to 2

(639) From 185 to 195

(640) From 134 to 4

Find the percent change of the below and state if it's a percentage increase or decrease. (Round to the nearest tenth)

(641) From 123 to 25

(642) From 58 to 28

(643) From 49 to 150

(644) From 150 to 123

(645) From 189 to 150

Find the percent change of the below and state if it's a percentage increase or decrease. (Round to the nearest tenth)

(646) From 143 to 148

(647) From 142 to 6

(648) From 119 to 16

(649) From 76 to 48

(650) From 67 to 127

Find the percent change of the below and state if it's a percentage increase or decrease. (Round to the nearest tenth)

(651) From 85 to 78

(652) From 46 to 74

(653) From 75 to 119

(654) From 123 to 74

(655) From 138 to 199

Find the percent change of the below and state if it's a percentage increase or decrease. (Round to the nearest tenth)

(656) From 52 to 155

(657) From 85 to 4

(658) From 172 to 58

(659) From 63 to 122

(660) From 25 to 97

Find the percent change of the below and state if it's a percentage increase or decrease. (Round to the nearest tenth)

(661) From $42 to $57

(662) From 50 m to 22 m

(663) From 27 miles to 75 miles

(664) From 72 minutes to 43 minutes

(665) From 86 m to 88 m

Find the percent change of the below and state if it's a percentage increase or decrease. (Round to the nearest tenth)

(666) From 67.8 minutes to 80 minutes

(667) From 46 miles to 73 miles

(668) From 8 miles to 25 miles

(669) From 95 miles to 41 miles

(670) From 1 minute to 4 minutes

Find the percent change of the below and state if it's a percentage increase or decrease. (Round to the nearest tenth)

(671) From 26 grams to 6 grams

(672) From $77 to $78.10

(673) From 13 hours to 15 hours

(674) From 23 tons to 21 tons

(675) From 98 minutes to 34 minutes

Find the percent change of the below and state if it's a percentage increase or decrease. (Round to the nearest tenth)

(676) From 64 ft to 6 ft

(677) From 12 inches to 30 inches

(678) From 84 hours to 60 hours

(679) From 56 tons to 83 tons

(680) From 54 miles to 45 miles

Find the percent change of the below and state if it's a percentage increase or decrease. (Round to the nearest tenth)

(681) From 23 inches to 1 inch

(682) From 52 inches to 44 inches

(683) From 39 minutes to 51 minutes

(684) From 27 tons to 6 tons

Find the percent change of the below and state if it's a percentage increase or decrease. (Round to the nearest tenth)

(685) From 58 m to 24 m

(686) From 7.3 tons to 27 tons

(687) From 85 grams to 48 grams

(688) From 45 m to 52 m

Find the percent change of the below and state if it's a percentage increase or decrease. (Round to the nearest tenth)

(689) From 62 inches to 77 inches

(690) From 36 inches to 59 inches

(691) From 9 miles to 21 miles

(692) From 80 m to 61 m

(693) From 25 tons to 44 tons

Find the percent change of the below and state if it's a percentage increase or decrease. (Round to the nearest tenth)

(694) From 79 ft to 62 ft

(695) From 61 hours to 74 hours

(696) From 100 grams to 32 grams

(697) From 71 tons to 88 tons

(698) From $85 to $24

Find the percent change of the below and state if it's a percentage increase or decrease. (Round to the nearest tenth)

(699) From 73 miles to 49 miles

(700) From 32 grams to 83 grams

(701) From 14 minutes to 35 minutes

(702) From 44 tons to 16 tons

(703) From $54 to $68

Find the percent change of the below and state if it's a percentage increase or decrease. (Round to the nearest tenth)

(704) From 31 hours to 38 hours

(705) From 92 grams to 63 grams

(706) From $95 to $78

(707) From $68 to $96

(708) From 51 ft to 58 ft

Find the percent change of the below and state if it's a percentage increase or decrease. (Round to the nearest tenth)

(709) From 43 tons to 49 tons

(710) From 56 grams to 47 grams

(711) From 97 miles to 57 miles

(712) From 76 tons to 28 tons

(713) From 53 hours to 84 hours

Find the percent change of the below and state if it's a percentage increase or decrease. (Round to the nearest tenth)

(714) From 14 m to 55 m

(715) From \$77 to \$84

(716) From 18 m to 86 m

(717) From 22 grams to 31 grams

(718) From 97 ft to 87 ft

Find the percent change of the below and state if it's a percentage increase or decrease. (Round to the nearest tenth)

(719) From 80 minutes to 29 minutes

(720) From 99 tons to 20 tons

(721) From $7 to $29

(722) From 38 ft to 60.1 ft

(723) From 44 ft to 35 ft

Find the percent change of the below and state if it's a percentage increase or decrease. (Round to the nearest tenth)

(724) From 69 grams to 29 grams

(725) From 34 m to 65 m

(726) From 18 ft to 37 ft

(727) From 18 tons to 61 tons

(728) From 92.6 hours to 46 hours

Find the percent change of the below and state if it's a percentage increase or decrease. (Round to the nearest tenth)

(729) From 47 minutes to 7 minutes

(730) From 58 miles to 9 miles

(731) From 79 grams to 10 grams

(732) From 95.7 ft to 3 ft

(733) From 100 minutes to 69 minutes

Find the percent change of the below and state if it's a percentage increase or decrease. (Round to the nearest tenth)

(734) From 69 tons to 83 tons

(735) From $87 to $88

(736) From 88 tons to 49 tons

(737) From $24 to $86

(738) From 91 tons to 55 tons

Find the percent change of the below and state if it's a percentage increase or decrease. (Round to the nearest tenth)

(739) From 55 m to 5 m

(740) From 27 m to 91 m

(741) The Original price of an item A is \$7.99. Store offers a discount of 20%. Find its selling price.

(742) The Original price of an item A is \$79.99. Store offers a discount of 30%. Find its selling price.

(743) The Original price of an item A is \$64.95. Store offers a discount of 30%. Find its selling price.

(744) The Original price of an item A is \$12.50. Store offers a discount of 20%. Find its selling price.

(745) The Original price of an item A is \$2.50. Store offers a discount of 40%. Find its selling price.

(746) The Original price of an item A is $24.95. Store offers a discount of 40%. Find its selling price.

(747) The Original price of an item A is $179.99. Store offers a discount of 16%. Find its selling price.

(748) The Original price of an item A is $199.99. Store offers a discount of 20%. Find its selling price.

(749) The Original price of an item A is $14.95. Store offers a discount of 31%. Find its selling price.

(750) The Original price of an item A is $449.95. Store offers a discount of 50%. Find its selling price.

(751) The Original price of an item A is $350.00. Store offers a discount of 50%. Find its selling price.

(752) The Original price of an item A is $0.99. Store offers a discount of 55%. Find its selling price.

(753) The Original price of an item A is $74.50. Store offers a discount of 30%. Find its selling price.

(754) The Original price of an item A is $67.95. Store offers a discount of 20%. Find its selling price.

(755) The Original price of an item A is $1,000.00. Store offers a discount of 50%. Find its selling price.

(756) The Original price of an item A is $139.99. Store offers a discount of 30%. Find its selling price.

(757) The Original price of an item A is $5,200.00. Store offers a discount of 27%. Find its selling price.

(758) The Original price of an item A is $139.99. Store offers a discount of 50%. Find its selling price.

(759) The Original price of an item A is $109.50. Store offers a discount of 38%. Find its selling price.

(760) The Original price of an item A is $119.99. Store offers a discount of 25%. Find its selling price.

(761) The Original price of an item A is $4.50. Store offers a discount of 29%. Find its selling price.

(762) The Original price of an item A is $149.95. Store offers a discount of 39%. Find its selling price.

(763) The Original price of an item A is 189.99. Store offers a discount of 20%. Find its selling price.

(764) The Original price of an item A is $40,500.00. Store offers a discount of 11%. Find its selling price.

(765) The Original price of an item A is $48.50. Store offers a discount of 40%. Find its selling price.

(766) The Original price of an item A is \$50.00. Store offers a discount of 45%. Find its selling price.

(767) The Original price of an item A is 150.00. Store offers a discount of 33%. Find its selling price.

(768) The Original price of an item A is \$155.00. Store offers a discount of 10%. Find its selling price.

(769) The Original price of an item A is \$60.99. Store offers a discount of 50%. Find its selling price.

(770) The Original price of an item A is \$18.99. Store offers a discount of 15%. Find its selling price.

(771) The Original price of an item A is \$19.99. Store offers a discount of 10%. Find its selling price.

(772) The Original price of an item A is \$64.95. Store offers a discount of 20%. Find its selling price.

(773) The Original price of an item A is \$25,000.00. Store offers a discount of 50%. Find its selling price.

(774) The Original price of an item A is \$199.50. Store offers a discount of 40%. Find its selling price.

(775) The Original price of an item A is \$74.99. Store offers a discount of 30%. Find its selling price.

(776) The Original price of an item A is $63.99. Store offers a discount of 47%. Find its selling price.

(777) The Original price of an item A is $4.50. Store offers a discount of 54%. Find its selling price.

(778) The Original price of an item A is $9.95. Store offers a discount of 20%. Find its selling price.

(779) The Original price of an item A is $17.99. Store offers a discount of 17%. Find its selling price.

(780) The Original price of an item A is $45.00. Store offers a discount of 9%. Find its selling price.

(781) The Original price of an item A is $39.9. Store offers a discount of 50%. Find its selling price.

(782) The Original price of an item A is $6.50. Store offers a discount of 40%. Find its selling price.

(783) The Original price of an item A is $549.50. Store offers a discount of 20%. Find its selling price.

(784) The Original price of an item A is $229.95. Store offers a discount of 25%. Find its selling price.

(785) The Original price of an item A is $69.95. Store offers a discount of 10%. Find its selling price.

(786) The Original price of an item A is \$689.50. Store offers a discount of 53%. Find its selling price.

(787) The Original price of an item A is \$25.00. Store offers a discount of 58%. Find its selling price.

(788) The Original price of an item A is \$4.99. Store offers a discount of 20%. Find its selling price.

(789) The Original price of an item A is \$69.95. Store offers a discount of 42%. Find its selling price.

(790) The Original price of an item A is \$249.50. Store offers a discount of 30%. Find its selling price.

(791) Original price of an item A is \$189.99. Store offers a discount of 45%. Find its selling price.

(792) The Original price of an item A is \$149.50. Store offers a discount of 50%. Find its selling price.

(793) The Original price of an item A is \$3,500.00. Store offers a discount of 40%. Find its selling price.

(794) The Original price of an item A is \$22.50. Store offers a discount of 45%. Find its selling price.

(795) The Original price of an item A is \$49.95. Store offers a discount of 30%. Find its selling price.

(796) The Original price of an item A is \$1.70. Store offers a discount of 40%. Find its selling price.

(797) The Original price of an item A is \$280.00. Store offers a discount of 5%. Find its selling price.

(798) The Original price of an item A is \$199.99. Store offers a discount of 60%. Find its selling price.

(799) The Original price of an item A is \$150.00. Store offers a discount of 20%. Find its selling price.

(800) The Original price of an item A is \$4.50. Store offers a discount of 5%. Find its selling price.

(801) The Original price of an item A is $124.99. Store offers a discount of 40%. Find its selling price.

(802) The Original price of an item A is $10,000.00. Store offers a discount of 40%. Find its selling price.

(803) The Original price of an item A is $31.50. Store offers a discount of 15%. Find its selling price.

(804) The Original price of an item A is $109.99. Store offers a discount of 11%. Find its selling price.

(805) The Original price of an item A is $109.95. Store offers a discount of 20%. Find its selling price.

(806) The Original price of an item A is $19.99. Store offers a discount of 20%. Find its selling price.

(807) The Original price of an item A is $94.95. Store offers a discount of 50%. Find its selling price.

(808) The Original price of an item A is $2.10. Store offers a discount of 15%. Find its selling price.

(809) The Original price of an item A is $31.50. Store offers a discount of 40%. Find its selling price.

(810) The Original price of an item A is $2.35. Store offers a discount of 50%. Find its selling price.

(811) The Original price of an item A is \$79.50. Store offers a discount of 30%. Find its selling price.

(812) The Original price of an item A is \$389.95. Store offers a discount of 20%. Find its selling price.

(813) The Original price of an item A is \$389.99. Store offers a discount of 26%. Find its selling price.

(814) The Original price of an item A is \$34.95. Store offers a discount of 20%. Find its selling price.

(815) The Original price of an item A is \$100.00. Store offers a discount of 50%. Find its selling price.

(816) Cathy purchased an Item at \$16.95 and sells it at a 30% price increase. Find the selling price of the item?

(817) Cathy purchased an Item at \$20.00 and sells it at a 90% price increase. Find the selling price of the item?

(818) Cathy purchased an Item at \$249.95 and sells it at a 85% price increase. Find the selling price of the item?

(819) Cathy purchased an Item at \$249.95 and sells it at a 79% price increase. Find the selling price of the item?

(820) Cathy purchased an Item at \$33.99 and sells it at a 20% price increase. Find the selling price of the item?

(821) Cathy purchased an Item at $49.50 and sells it at a 10% price increase. Find the selling price of the item?

(822) Cathy purchased an Item at $1,200.00 and sells it at a 30% price increase. Find the selling price of the item?

(823) Cathy purchased an Item at $15,000.00 and sells it at a 70% price increase. Find the selling price of the item?

(824) Cathy purchased an Item at $14,400.00 and sells it at a 10% price increase. Find the selling price of the item?

(825) Cathy purchased an Item at $40.00 and sells it at a 35% price increase. Find the selling price of the item?

(826) Cathy purchased an Item at $40,000.00 and sells it at a 80% price increase. Find the selling price of the item?

(827) Cathy purchased an Item at $38,000.00 and sells it at a 45% price increase. Find the selling price of the item?

(828) Cathy purchased an Item at $44.95 and sells it at a 10% price increase. Find the selling price of the item?

(829) Cathy purchased an Item at $45,000.00 and sells it at a 10% price increase. Find the selling price of the item?

(830) Cathy purchased an Item at $64.99 and sells it at a 70% price increase. Find the selling price of the item?

(831) Cathy purchased an Item at $27.99 and sells it at a 48% price increase. Find the selling price of the item?

(832) Cathy purchased an Item at $21,000.00 and sells it at a 10% price increase. Find the selling price of the item?

(833) Cathy purchased an Item at $979.50 and sells it at a 75% price increase. Find the selling price of the item?

(834) Cathy purchased an Item at $41.99 and sells it at a 20% price increase. Find the selling price of the item?

(835) Cathy purchased an Item at $149.50 and sells it at a 10% price increase. Find the selling price of the item?

(836) Cathy purchased an Item at $54.50 and sells it at a 50% price increase. Find the selling price of the item?

(837) Cathy purchased an Item at $299.99 and sells it at a 30% price increase. Find the selling price of the item?

(838) Cathy purchased an Item at $14.95 and sells it at a 40% price increase. Find the selling price of the item?

(839) Cathy purchased an Item at $35.95 and sells it at a 70% price increase. Find the selling price of the item?

(840) Cathy purchased an Item at $31.50 and sells it at a 70% price increase. Find the selling price of the item?

(841) Cathy purchased an Item at $2,400.00 and sells it at a 35% price increase. Find the selling price of the item?

(842) Cathy purchased an Item at $34.95 and sells it at a 50% price increase. Find the selling price of the item?

(843) Cathy purchased an Item at $22.99 and sells it at a 49% price increase. Find the selling price of the item?

(844) Cathy purchased an Item at $1,800.00 and sells it at a 50% price increase. Find the selling price of the item?

(845) Cathy purchased an Item at $14.50 and sells it at a 35% price increase. Find the selling price of the item?

(846) Cathy purchased an Item at $124.99 and sells it at a 85% price increase. Find the selling price of the item?

(847) Cathy purchased an Item at $119.99 and sells it at a 50% price increase. Find the selling price of the item?

(848) Cathy purchased an Item at $1.99 and sells it at a 50% price increase. Find the selling price of the item?

(849) Cathy purchased an Item at $119.50 and sells it at a 45% price increase. Find the selling price of the item?

(850) Cathy purchased an Item at $39.50 and sells it at a 58% price increase. Find the selling price of the item?

(851) Cathy purchased an Item at \$249.95 and sells it at a 24% price increase. Find the selling price of the item?

(852) Cathy purchased an Item at \$79.99 and sells it at a 50% price increase. Find the selling price of the item?

(853) Cathy purchased an Item at \$5.99 and sells it at a 43% price increase. Find the selling price of the item?

(854) Cathy purchased an Item at \$129.95 and sells it at a 10% price increase. Find the selling price of the item?

(855) Cathy purchased an Item at \$30,000.00 and sells it at a 65% price increase. Find the selling price of the item?

(856) Cathy purchased an Item at $4.95 and sells it at a 40% price increase. Find the selling price of the item?

(857) Cathy purchased an Item at $2,850.00 and sells it at a 18% price increase. Find the selling price of the item?

(858) Cathy purchased an Item at $6.00 and sells it at a 40% price increase. Find the selling price of the item?

(859) Cathy purchased an Item at $75.50 and sells it at a 75% price increase. Find the selling price of the item?

(860) Cathy purchased an Item at $109.99 and sells it at a 50% price increase. Find the selling price of the item?

(861) Cathy purchased an Item at $34.50 and sells it at a 20% price increase. Find the selling price of the item?

(862) Cathy purchased an Item at $8.95 and sells it at a 60% price increase. Find the selling price of the item?

(863) Cathy purchased an Item at $29.50 and sells it at a 35% price increase. Find the selling price of the item?

(864) Cathy purchased an Item at $2.50 and sells it at a 10% price increase. Find the selling price of the item?

(865) Cathy purchased an Item at $44.95 and sells it at a 90% price increase. Find the selling price of the item?

(866) Cathy purchased an Item at \$19.00 and sells it at a 75% price increase. Find the selling price of the item?

(867) Cathy purchased an Item at \$2.50 and sells it at a 20% price increase. Find the selling price of the item?

(868) Cathy purchased an Item at \$149.99 and sells it at a 5% price increase. Find the selling price of the item?

(869) Cathy purchased an Item at \$11.95 and sells it at a 10% price increase. Find the selling price of the item?

(870) Cathy purchased an Item at \$9.50 and sells it at a 80% price increase. Find the selling price of the item?

(871) Cathy purchased an Item at $370.00 and sells it at a 65% price increase. Find the selling price of the item?

(872) Cathy purchased an Item at $5.95 and sells it at a 30% price increase. Find the selling price of the item?

(873) Cathy purchased an Item at $249.95 and sells it at a 60 %price increase. Find the selling price of the item?

(874) Cathy purchased an Item at $260.00 and sells it at a 40% price increase. Find the selling price of the item?

(875) Cathy purchased an Item at $19.95 and sells it at a 30% price increase. Find the selling price of the item?

(876) Cathy purchased an Item at $19.00 and sells it at a 20% price increase. Find the selling price of the item?

(877) Cathy purchased an Item at $150.00 and sells it at a 70% price increase. Find the selling price of the item?

(878) Cathy purchased an Item at $99.50 and sells it at a 20% price increase. Find the selling price of the item?

(879) Cathy purchased an Item at $44.95 and sells it at a 45% price increase. Find the selling price of the item?

(880) Cathy purchased an Item at $21.50 and sells it at a 35% price increase. Find the selling price of the item?

(881) Cathy purchased an Item at $67.99 and sells it at a 40% price increase. Find the selling price of the item?

(882) Cathy purchased an Item at $49.99 and sells it at a 20% price increase. Find the selling price of the item?

(883) Cathy purchased an Item at $9.50 and sells it at a 74% price increase. Find the selling price of the item?

(884) Cathy purchased an Item at $149.50 and sells it at a 45% price increase. Find the selling price of the item?

(885) Cathy purchased an Item at $2,000.00 and sells it at a 50% price increase. Find the selling price of the item?

(886) Cathy purchased an Item at $14.50 and sells it at a 75% price increase. Find the selling price of the item?

(887) Cathy purchased an Item at $4.00 and sells it at a 93% price increase. Find the selling price of the item?

(888) Cathy purchased an Item at $6,700.00 and sells it at a 90% price increase. Find the selling price of the item?

(889) Cathy purchased an Item at $8,800.00 and sells it at a 45% price increase. Find the selling price of the item?

(890) Cathy purchased an Item at $47,000.00 and sells it at a 40% price increase. Find the selling price of the item?

(891) The original price of an Item X is $134.95 and a tax of 3% is charged on the sale. Find the selling price of the Item X

(892) The original price of an Item X is $22.50 and a tax of 4% is charged on the sale. Find the selling price of the Item X

(893) The original price of an Item X is $17.50 and a tax of 2% is charged on the sale. Find the selling price of the Item X

(894) The original price of an Item X is $56.00 and a tax of 1% is charged on the sale. Find the selling price of the Item X

(895) The original price of an Item X is $44.95 and a tax of 1% is charged on the sale. Find the selling price of the Item X

(896) The original price of an Item X is $9.50 and a tax of 2% is charged on the sale. Find the selling price of the Item X

(897) The original price of an Item X is $16.50 and a tax of 5% is charged on the sale. Find the selling price of the Item X

(898) The original price of an Item X is $74.95 and a tax of 3% is charged on the sale. Find the selling price of the Item X

(899) The original price of an Item X is $29.50 and a tax of 1% is charged on the sale. Find the selling price of the Item X

(900) The original price of an Item X is $20.00 and a tax of 4% is charged on the sale . Find the selling price of the Item X

(901) The original price of an Item X is $2.50 and a tax of 2% is charged on the sale. Find the selling price of the Item X

(902) The original price of an Item X is $89.50 and a tax of 4% is charged on the sale. Find the selling price of the Item X

(903) The original price of an Item X is $150.00 and a tax of 6% is charged on the sale. Find the selling price of the Item X

(904) The original price of an Item X is $170.00 and a tax of 4% is charged on the sale. Find the selling price of the Item X

(905) The original price of an Item X is $149.50 and a tax of 4% is charged on the sale. Find the selling price of the Item X

(906) The original price of an Item X is $19.95 and a tax of 1% is charged on the sale. Find the selling price of the Item X

(907) The original price of an Item X is $10.99 and a tax of 5% is charged on the sale. Find the selling price of the Item X

(908) The original price of an Item X is $8.50 and a tax of 3% is charged on the sale. Find the selling price of the Item X

(909) The original price of an Item X is $420.00 and a tax of 2% is charged on the sale. Find the selling price of the Item X

(910) The original price of an Item X is $2.50 and a tax of 1% is charged on the sale. Find the selling price of the Item X

(911) The original price of an Item X is \$290.00 and a tax of 6% is charged on the sale. Find the selling price of the Item X

(912) The original price of an Item X is \$34.99 and a tax of 5% is charged on the sale. Find the selling price of the Item X

(913) The original price of an Item X is \$239.50 and a tax of 6% is charged on the sale. Find the selling price of the Item X

(914) The original price of an Item X is \$13.95 and a tax of 6% is charged on the sale. Find the selling price of the Item X

(915) The original price of an Item X is \$59.50 and a tax of 6% is charged on the sale. Find the selling price of the Item X

(916) The original price of an Item X is $159.95 and a tax of 5% is charged on the sale. Find the selling price of the Item X

(917) The original price of an Item X is $5,100.00 and a tax of 5% is charged on the sale. Find the selling price of the Item X

(918) The original price of an Item X is $379.95 and a tax of 4% is charged on the sale. Find the selling price of the Item X

(919) The original price of an Item X is $76.00 and a tax of 4% is charged on the sale. Find the selling price of the Item X

(920) The original price of an Item X is $34,500.00 and a tax of 6% is charged on the sale. Find the selling price of the Item X

(921) The original price of an Item X is $19.99 and a tax of 2% is charged on the sale. Find the selling price of the Item X

(922) The original price of an Item X is $9.50 and a tax of 6% is charged on the sale. Find the selling price of the Item X

(923) The original price of an Item X is $14.50 and a tax of 5% is charged on the sale. Find the selling price of the Item X

(924) The original price of an Item X is $3.95 and a tax of 5% is charged on the sale. Find the selling price of the Item X

(925) The original price of an Item X is $4.00 and a tax of 6% is charged on the sale. Find the selling price of the Item X

(926) The original price of an Item X is $149.99 and a tax of 4% is charged on the sale. Find the selling price of the Item X

(927) The original price of an Item X is $29.95 and a tax of 4% is charged on the sale. Find the selling price of the Item X

(928) The original price of an Item X is $49.50 and a tax of 6% is charged on the sale. Find the selling price of the Item X

(929) The original price of an Item X is $10.99 and a tax of 3% is charged 9 on the sale. Find the selling price of the Item X

(930) The original price of an Item X is $2,700.00 and a tax of 6% is charged on the sale. Find the selling price of the Item X

(931) The original price of an Item X is $47.00 and a tax of 5% is charged on the sale. Find the selling price of the Item X

(932) The original price of an Item X is $414.95 and a tax of 2% is charged on the sale. Find the selling price of the Item X

(933) The original price of an Item X is $1.95 and a tax of 2% is charged on the sale. Find the selling price of the Item X

(934) The original price of an Item X is $84.50 and a tax of 3% is charged on the sale. Find the selling price of the Item X

(935) The original price of an Item X is $89.95 and a tax of 2% is charged on the sale. Find the selling price of the Item X

(936) The original price of an Item X is $99.95 and a tax of 6% is charged on the sale. Find the selling price of the Item X

(937) The original price of an Item X is $49.50 and a tax of 4% is charged on the sale. Find the selling price of the Item X

(938) The original price of an Item X is $18.99 and a tax of 4% is charged on the sale. Find the selling price of the Item X

(939) The original price of an Item X is $50.00 and a tax of 3% is charged on the sale. Find the selling price of the Item X

(940) The original price of an Item X is $299.50 and a tax of 5% is charged on the sale. Find the selling price of the Item X

(941) The original price of an Item X is \$1,400.00 and a tax of 5% is charged on the sale. Find the selling price of the Item X

(942) The original price of an Item X is \$24.99 and a tax of 5% is charged on the sale. Find the selling price of the Item X

(943) The original price of an Item X is \$400.00 and a tax of 1% is charged on the sale. Find the selling price of the Item X

(944) The original price of an Item X is \$16,000.00 and a tax of 5% is charged on the sale. Find the selling price of the Item X

(945) The original price of an Item X is \$25.50 and a tax of 1% is charged on the sale. Find the selling price of the Item X

(946) The original price of an Item X is \$620.00 and a tax of 4% is charged on the sale. Find the selling price of the Item X

(947) The original price of an Item X is \$39.95 and a tax of 6% is charged on the sale. Find the selling price of the Item X

(948) The original price of an Item X is \$199.95 and a tax of 4% is charged on the sale. Find the selling price of the Item X

(949) The original price of an Item X is \$300.00 and a tax of 2% is charged on the sale. Find the selling price of the Item X

(950) The original price of an Item X is \$33,000.00 and a tax of 1% is charged on the sale. Find the selling price of the Item X

(951) The original price of an Item X is $44.95 and a tax of 3% is charged on the sale. Find the selling price of the Item X

(952) The original price of an Item X is $2.50 and a tax of 3% is charged on the sale. Find the selling price of the Item X

(953) The original price of an Item X is $99.99 and a tax of 2% is charged on the sale. Find the selling price of the Item X

(954) The original price of an Item X is $29.99 and a tax of 5% is charged on the sale. Find the selling price of the Item X

(955) The original price of an Item X is $8.99 and a tax of 3% is charged on the sale. Find the selling price of the Item X

(956) The original price of an Item X is $249.99 and a tax of 4% is charged on the sale. Find the selling price of the Item X

(957) The original price of an Item X is $110.00 and a tax of 3% is charged on the sale. Find the selling price of the Item X

(958) The original price of an Item X is $0.99 and a tax of 2% is charged on the sale. Find the selling price of the Item X

(959) The original price of an Item X is $820.00 and a tax of 2% is charged on the sale. Find the selling price of the Item X

(960) The original price of an Item X is $23,500.00 and a tax of 3% is charged on the sale. Find the selling price of the Item X

(961) The original price of an Item X is $480.00 and a tax of 3% is charged on the sale. Find the selling price of the Item X

(962) The original price of an Item X is $59.50 and a tax of 2% is charged on the sale. Find the selling price of the Item X

(963) The original price of an Item X is $64.99 and a tax of 3% is charged on the sale. Find the selling price of the Item X

(964) The original price of an Item X is $31.00 and a tax of 4% is charged on the sale. Find the selling price of the Item X

(965) The original price of an Item X is $19.50 and a tax of 5% is charged on the sale. Find the selling price of the Item X

(966) A simple interest is calculated on a loan amount of $21,900 at 8% for 2 years. Find the ending balance.

(967) A simple interest is calculated on a loan amount of $1,670 at 5% for 4 years. Find the ending balance.

(968) A simple interest is calculated on a loan amount of $210 at 16% for 2 years. Find the ending balance.

(969) A simple interest is calculated on a loan amount of $240 at 10% for 4 years. Find the ending balance.

(970) A simple interest is calculated on a loan amount of $450 at 9% for 6 years. Find the ending balance.

(971) A simple interest is calculated on a loan amount of $680 at 13% for 2 years. Find the ending balance.

(972) A simple interest is calculated on a loan amount of $10,000 at 14% for 4 years. Find the ending balance.

(973) A simple interest is calculated on a loan amount of $1,100 at 13% for 3 years. Find the ending balance.

(974) A simple interest is calculated on a loan amount of $6,300 at 7% for 4 years. Find the ending balance.

(975) A simple interest is calculated on a loan amount of $27,900 at 10% for 3 years. Find the ending balance.

(976) A simple interest is calculated on a loan amount of $350 at 15% for 4 years. Find the ending balance.

(977) A simple interest is calculated on a loan amount of $12,300 at 4% for 2 years. Find the ending balance.

(978) A simple interest is calculated on a loan amount of $46,400 at 2% for 3 years. Find the ending balance.

(979) A simple interest is calculated on a loan amount of $33,000 at 2% for 2 years. Find the ending balance.

(980) A simple interest is calculated on a loan amount of $30,000 at 3% for 4 years. Find the ending balance.

(981) A simple interest is calculated on a loan amount of $220 at 9% for 3 years. Find the ending balance.

(982) A simple interest is calculated on a loan amount of $1,850 at 16% for 5 years. Find the ending balance.

(983) A simple interest is calculated on a loan amount of $4,000 at 2% for 3 years. Find the ending balance.

(984) A simple interest is calculated on a loan amount of $340 at 9% for 5 years. Find the ending balance.

(985) A simple interest is calculated on a loan amount of $550 at 5% for 4 years. Find the ending balance.

(986) A simple interest is calculated on a loan amount of $70 at 14% for 2 years. Find the ending balance.

(987) A simple interest is calculated on a loan amount of $1,070 at 1% for 6 years. Find the ending balance.

(988) A simple interest is calculated on a loan amount of $39,000 at 10% for 3 years. Find the ending balance.

(989) A simple interest is calculated on a loan amount of $45,000 at 8% for 2 years. Find the ending balance.

(990) A simple interest is calculated on a loan amount of $510 at 1% for 2 years. Find the ending balance.

(991) A simple interest is calculated on a loan amount of $120 at 8% for 5 years. Find the ending balance.

(992) A simple interest is calculated on a loan amount of $2,000 at 3% for 7 years. Find the ending balance.

(993) A simple interest is calculated on a loan amount of $640 at 10% for 3 years. Find the ending balance.

(994) A simple interest is calculated on a loan amount of $1,240 at 8% for 4 years. Find the ending balance.

(995) A simple interest is calculated on a loan amount of $135 at 9% for 5 years. Find the ending balance.

(996) A simple interest is calculated on a loan amount of $28,100 at 3% for 9 years. Find the ending balance.

(997) A simple interest is calculated on a loan amount of $1,000 at 13% for 2 years. Find the ending balance.

(998) A simple interest is calculated on a loan amount of $46,200 at 9% for 8 years. Find the ending balance.

(999) A simple interest is calculated on a loan amount of $370 at 3% for 3 years. Find the ending balance.

(1000) A simple interest is calculated on a loan amount of $11,200 at 6% for 5 years. Find the ending balance.

(1001) A simple interest is calculated on a loan amount of $405 at 13% for 2 years. Find the ending balance.

(1002) A simple interest is calculated on a loan amount of $55,000 at 7% for 4 years. Find the ending balance.

(1003) A simple interest is calculated on a loan amount of $12,000 at 8% for 2 years. Find the ending balance.

(1004) A simple interest is calculated on a loan amount of $200 at 13% fo 3 years. Find the ending balance.

(1005) A simple interest is calculated on a loan amount of $27,000 at 8% for 2 years. Find the ending balance.

(1006) A simple interest is calculated on a loan amount of $1,200 at 6% for 8 years. Find the ending balance.

(1007) A simple interest is calculated on a loan amount of $55,600 at 13% for 6 years. Find the ending balance.

(1008) A simple interest is calculated on a loan amount of $40 at 4% for 2 years. Find the ending balance.

(1009) A simple interest is calculated on a loan amount of $430 at 15% for 2 years. Find the ending balance.

(1010) A simple interest is calculated on a loan amount of $4,100 at 5% for 6 years. Find the ending balance.

(1011) A simple interest is calculated on a loan amount of $43,000 at 16% for 4 years. Find the ending balance.

(1012) A simple interest is calculated on a loan amount of $420 at 1% for 4 years. Find the ending balance.

(1013) A simple interest is calculated on a loan amount of $870 at 8% for 2 years. Find the ending balance.

(1014) A simple interest is calculated on a loan amount of $950 at 5% for 3 years. Find the ending balance.

(1015) A simple interest is calculated on a loan amount of $24,500 at 1% for 2 years. Find the ending balance.

(1016) A simple interest is calculated on a loan amount of $32,000 at 3% for 3 years. Find the ending balance.

(1017) A simple interest is calculated on a loan amount of $39,000 at 12% for 2 years. Find the ending balance.

(1018) A simple interest is calculated on a loan amount of $31,000 at 7% for 2 years. Find the ending balance.

(1019) A simple interest is calculated on a loan amount of $42,700 at 2% for 9 years. Find the ending balance.

(1020) A simple interest is calculated on a loan amount of $57,200 at 3% for 5 years. Find the ending balance.

(1021) A simple interest is calculated on a loan amount of $17,300 at 8% for 4 years. Find the ending balance.

(1022) A simple interest is calculated on a loan amount of $8,000 at 5% for 2 years. Find the ending balance.

(1023) A simple interest is calculated on a loan amount of $25,000 at 16% for 2 years. Find the ending balance.

(1024) A simple interest is calculated on a loan amount of $18,100 at 7% for 3 years. Find the ending balance.

(1025) A simple interest is calculated on a loan amount of $7,500 at 13% for 2 years. Find the ending balance.

Find the total value of the investment after the time given.

(1026) A compound interest is calculated on a loan amount of $9,000 at 11% compounded annually for 2 years. Find the ending balance of the loan.

(1027) A compound interest is calculated on a loan amount of $390 at 11% compounded annually for 2 years. Find the ending balance of the loan.

(1028) A compound interest is calculated on a amount of $1,150 at 14% compounded annually for 6 years. Find the ending balance of the loan.

(1029) A compound interest is calculated on a loan amount of $46,700 at 3% compounded annually for 5 years. Find the ending balance of the loan.

(1030) A compound interest is calculated on a loan amount of $43,000 at 6% compounded annually for 9 years. Find the ending balance of the loan.

(1031) A compound interest is calculated on a loan amount of $54,900 at 11% compounded annually for 2 years. Find the ending balance of the loan.

(1032) A compound interest is calculated on a loan amount of $275 at 11% compounded annually for 4 years. Find the ending balance of the loan.

(1033) A compound interest is calculated on a loan amount of $900 at 11% compounded annually for 5 years. Find the ending balance of the loan.

(1034) A compound interest is calculated on a loan amount of $210 at 11% compounded annually for 2 years. Find the ending balance of the loan.

(1035) A compound interest is calculated on a loan amount of $7,100 at 12% compounded annually for 4 years. Find the ending balance of the loan.

(1036) A compound interest is calculated on a loan amount of $51,700 at 14% ompounded annually for 2 years. Find the ending balance of the loan.

(1037) A compound interest is calculated on a loan amount of $59,300 at 2% compounded annually for 3 years. Find the ending balance of the loan.

(1038) A compound interest is calculated on a loan amount of $1,840 at 10% compounded annually for 3 years. Find the ending balance of the loan.

(1039) A compound interest is calculated on a loan amount of $235 at 10% compounded annually for 7 years. Find the ending balance of the loan.

(1040) A compound interest is calculated on a loan amount of $730 at 10% compounded annually for 2 years. Find the ending balance of the loan.

(1041) A compound interest is calculated on a loan amount of $5,000 at 2% compounded annually for 2 years. Find the ending balance of the loan.

(1042) A compound interest is calculated on a loan amount of $57,900 at 13% compounded annually for 7 years. Find the ending balance of the loan.

(1043) A compound interest is calculated on a loan amount of $1,600 at 1% compounded annually for 3 years. Find the ending balance of the loan.

(1044) A compound interest is calculated on a loan amount of $170 at 15% compounded annually for 3 years. Find the ending balance of the loan.

(1045) A compound interest is calculated on a loan amount of $1,210 at 10% compounded annually for 4 years. Find the ending balance of the loan.

(1046) A compound interest is calculated on a loan of $640 at 16% compounded annually for 2 years. Find the ending balance of the loan.

(1047) A compound interest is calculated on a loan amount of $54,100 at 6% compounded annually for 2 years. Find the ending balance of the loan.

(1048) A compound interest is calculated on a loan amount of $51,400 at 12% compounded annually for 4 years. Find the ending balance of the loan.

(1049) A compound interest is calculated on a loan amount of $1,270 at 2% compounded annually for 2 years. Find the ending balance of the loan.

(1050) A compound interest is calculated on a loan amount of $1,850 at 5% compounded annually for 5 years. Find the ending balance of the loan.

(1051) A compound interest is calculated on a loan amount of $23,300 at 12% compounded annually for 8 years. Find the ending balance of the loan.

(1052) A compound interest is calculated on a loan amount of $57,000 at 7% compounded annually for 2 years. Find the ending balance of the loan.

(1053) A compound interest is calculated on a loan amount of $205 at 13% compounded annually for 6 years. Find the ending balance of the loan.

(1054) A compound interest is calculated on a loan amount of $59,000 at 12% compounded annually for 6 years. Find the ending balance of the loan.

(1055) A compound interest is calculated on a loan amount of $590 at 11% compounded annually for 8 years. Find the ending balance of the loan.

(1056) A compound interest is calculated on a loan amount of $12,000 at 13% compounded annually for 2 years. Find the ending balance of the loan.

(1057) A compound interest is calculated on a loan amount of $260 at 9% compounded annually for 2 years. Find the ending balance of the loan.

(1058) A compound interest is calculated on a loan amount of $25 at 3% compounded annually for 2 years. Find the ending balance of the loan.

(1059) A compound interest is calculated on a loan amount of $6,000 at 9% compounded annually for 10 years. Find the ending balance of the loan.

(1060) A compound interest is calculated on a loan amount of $27,000 at 13% compounded annually for 2 years. Find the ending balance of the loan.

(1061) A compound interest is calculated on a loan amount of $475 at 10% compounded annually for 3 years. Find the ending balance of the loan.

(1062) A compound interest is calculated on a loan amount of $7,000 at 4% compounded annually for 2 years. Find the ending balance of the loan.

(1063) A compound interest is calculated on a loan amount of $610 at 2% compounded annually for 2 years. Find the ending balance of the loan.

(1064) A compound interest is calculated on a loan amount of $1,400 at 3% compounded annually for4 years. Find the ending balance of the loan.

(1065) A compound interest is calculated on a loan amount of $18,500 at 7% compounded annually for 2 years. Find the ending balance of the loan.

(1066) A compound interest is calculated on a loan amount of \$58,000 at 2% compounded annually for 4 years. Find the ending balance of the loan.

(1067) A compound interest is calculated on a loan amount of \$55,900 at 6% compounded annually for 7 years. Find the ending balance of the loan.

(1068) A compound interest is calculated on a loan amount of \$470 at 16% compounded annually for 4 years. Find the ending balance of the loan.

(1069) A compound interest is calculated on a loan amount of \$36,000 at 6% compounded annually for 8 years. Find the ending balance of the loan.

(1070) A compound interest is calculated on a loan amount of \$21,000 at 12% compounded annually for 3 years. Find the ending balance of the loan.

(1071) A compound interest is calculated on a loan amount of $39,600 at 16% compounded annually for 7 years. Find the ending balance of the loan.

(1072) A compound interest is calculated on a loan amount of $1,390 at 9% compounded annually for 2 years. Find the ending balance of the loan.

(1073) A compound interest is calculated on a loan amount of $1,560 at 8% compounded annually for 2 years. Find the ending balance of the loan.

(1074) A compound interest is calculated on a loan amount of $130 at 8% compounded annually for 2 years. Find the ending balance of the loan.

(1075) A compound interest is calculated on a loan amount of $650 at 5% compounded annually for 4 years. Find the ending balance of the loan.

(1076) A compound interest is calculated on a loan amount of $415 at 10% compounded annually for 5 years. Find the ending balance of the loan.

(1077) A compound interest is calculated on a loan amount of $1,570 at 8% compounded annually for 9 years. Find the ending balance of the loan.

(1078) A compound interest is calculated on a loan amount of $52,100 at 10% compounded annually for 2 years. Find the ending balance of the loan.

(1079) A compound interest is calculated on a loan amount of $32,000 at 8% compounded annually for 10 years. Find the ending balance of the loan.

(1080) A compound interest is calculated on a loan amount of $1,920 at 12% compounded annually for 3 years. Find the ending balance of the loan. s

(1081) A compound interest is calculated on a loan amount of $470 at 15% compounded annually for 4 years. Find the ending balance of the loan.

(1082) A compound interest is calculated on a loan amount of $130 at 11% compounded annually for 2 years. Find the ending balance of the loan.

(1083) A compound interest is calculated on a loan amount of $50,700 at 4% compounded annually for 2 years. Find the ending balance of the loan.

(1084) A compound interest is calculated on a loan amount of $12,000 at 4% compounded annually for 7 years. Find the ending balance of the loan.

(1085) A compound interest is calculated on a loan amount of $58,000 at 15% compounded annually for 4 years. Find the ending balance of the loan.

(1086) A compound interest is calculated on a loan amount of $25,500 at 13% compounded semiannually 3 years. Find the ending balance of the loan.

(1087) A compound interest is calculated on a loan amount of $175 at 3% compounded semiannually 5 years. Find the ending balance of the loan.

(1088) A compound interest is calculated on a loan amount of $840 at 10% compounded semiannually 9 years. Find the ending balance of the loan.

(1089) A compound interest is calculated on a loan amount of $58,900 at 14% compounded semiannually 1 year. Find the ending balance of the loan.

(1090) A compound interest is calculated on a loan amount of $42,000 at 3% compounded semiannually 2 years. Find the ending balance of the loan.

(1091) A compound interest is calculated on a loan amount of $920 at 13% compounded semiannually 2 years . Find the ending balance of the loan.

(1092) A compound interest is calculated on a loan amount of $25,000 at 2% compounded semiannually 6 years. Find the ending balance of the loan.

(1093) A compound interest is calculated on a loan amount of $910 at 4% compounded semiannually 5 years. Find the ending balance of the loan.

(1094) A compound interest is calculated on a loan amount of $260 at 14% compounded semiannually 6 years . Find the ending balance of the loan.

(1095) A compound interest is calculated on a loan amount of $36,700 at 16% compounded semiannually 2 years. Find the ending balance of the loan.

(1096) A compound interest is calculated on a loan amount of $41,700 at 8% compounded semiannually 7 years. Find the ending balance of the loan.

(1097) A compound interest is calculated on a loan amount of $9,000 at 16% compounded semiannually 8 years. Find the ending balance of the loan.

(1098) A compound interest is calculated on a loan amount of $3,500 at 13% compounded semiannually 2 years. Find the ending balance of the loan.

(1099) A compound interest is calculated on a loan amount of $100 at 2% compounded semiannually 1 year. Find the ending balance of the loan.

(1100) A compound interest is calculated on a loan amount of $46,000 at 10% compounded semiannually 1 year. Find the ending balance of the loan.

(1101) A compound interest is calculated on a loan amount of $4,900 at 15% compounded semiannually 6 years. Find the ending balance of the loan.

(1102) A compound interest is calculated on a loan amount of $15,200 at 1% compounded semiannually 3 years. Find the ending balance of the loan.

(1103) A compound interest is calculated on a loan amount of $48,000 at 14% compounded semiannually 5 years. Find the ending balance of the loan.

(1104) A compound interest is calculated on a loan amount of $260 at 8% compounded semiannually 2 years. Find the ending balance of the loan.

(1105) A compound interest is calculated on a loan amount of $350 at 10% compounded semiannually 2 years. Find the ending balance of the loan.

(1106) A compound interest is calculated on a loan amount of $54,900 at 7% compounded semiannually 3 years. Find the ending balance of the loan.

(1107) A compound interest is calculated on a loan amount of $1,220 at 16% compounded semiannually 2 years . Find the ending balance of the loan.

(1108) A compound interest is calculated on a loan amount of $15,700 at 13% compounded semiannually 2 years. Find the ending balance of the loan.

(1109) A compound interest is calculated on a loan amount of $6,300 at 1% compounded semiannually 3 years. Find the ending balance of the loan.

(1110) A compound interest is calculated on a loan amount of $41,900 at 15% compounded semiannually 7 years. Find the ending balance of the loan.

(1111) A compound interest is calculated on a loan amount of $40,400 at 9% compounded semiannually 2 years. Find the ending balance of the loan.

(1112) A compound interest is calculated on a loan amount of $48,000 at 2% compounded semiannually 7 years. Find the ending balance of the loan.

(1113) A compound interest is calculated on a loan amount of $640 at 4% compounded semiannually 2 years. Find the ending balance of the loan.

(1114) A compound interest is calculated on a loan amount of $830 at 8% compounded semiannually 2 years. Find the ending balance of the loan.

(1115) A compound interest is calculated on a loan amount of $1,000 at 1% compounded semiannually 4 years. Find the ending balance of the loan.

(1116) A compound interest is calculated on a loan amount of $1,770 at 1% compounded semiannually 7 years. Find the ending balance of the loan.

(1117) A compound interest is calculated on a loan amount of $2,000 at 14% compounded semiannually 7 years. Find the ending balance of the loan.

(1118) A compound interest is calculated on a loan amount of $195 at 7% compounded semiannually 2 years. Find the ending balance of the loan.

(1119) A compound interest is calculated on a loan amount of $29,000 at 9% compounded semiannually 1 year. Find the ending balance of the loan.

(1120) A compound interest is calculated on a loan amount of $15,000 at 15% compounded semiannually 3 years. Find the ending balance of the loan.

(1121) A compound interest is calculated on a loan amount of $980 at 7% compounded semiannually 4 years. Find the ending balance of the loan.

(1122) A compound interest is calculated on a loan amount of $49,000 at 8% compounded semiannually 2 years. Find the ending balance of the loan.

(1123) A compound interest is calculated on a loan amount of $26,400 at 13% compounded semiannually 4 years. Find the ending balance of the loan.

(1124) A compound interest is calculated on a loan amount of $5,200 at 6% compounded semiannually 2 years. Find the ending balance of the loan.

(1125) A compound interest is calculated on a loan amount of $710 at 7% compounded semiannually 7 years. Find the ending balance of the loan.

(1126) A compound interest is calculated on a loan amount of $32,500 at 2% compounded semiannually 1 year. Find the ending balance of the loan.

(1127) A compound interest is calculated on a loan amount of $1,900 at 10% compounded semiannually 7 years. Find the ending balance of the loan.

(1128) A compound interest is calculated on a loan amount of $330 at 5% compounded semiannually 3 years. Find the ending balance of the loan.

(1129) A compound interest is calculated on a loan amount of $10,400 at 1% compounded semiannually 6 years. Find the ending balance of the loan.

(1130) A compound interest is calculated on a loan amount of $100 at 11% compounded semiannually 8 years. Find the ending balance of the loan.

(1131) A compound interest is calculated on a loan amount of $26,000 at 3% compounded semiannually 2 years. Find the ending balance of the loan.

(1132) A compound interest is calculated on a loan amount of $1,790 at 11% compounded semiannually 2 years. Find the ending balance of the loan.

(1133) A compound interest is calculated on a loan amount of $24,900 at 13% compounded semiannually 2 years. Find the ending balance of the loan.

(1134) A compound interest is calculated on a loan amount of $4,000 at 3% compounded semiannually 5 years. Find the ending balance of the loan.

(1135) A compound interest is calculated on a loan amount of $135 at 1% compounded semiannually 1 years. Find the ending balance of the loan.

(1136) A compound interest is calculated on a loan amount of $31,000 at 9% compounded semiannually 6 years. Find the ending balance of the loan.

(1137) A compound interest is calculated on a loan amount of $1,710 at 5% compounded semiannually 2 years. Find the ending balance of the loan.

(1138) A compound interest is calculated on a loan amount of $1,100 at 2% compounded semiannually 2 years. Find the ending balance of the loan.

(1139) A compound interest is calculated on a loan amount of $170 at 2% compounded semiannually 6 years. Find the ending balance of the loan.

(1140) A compound interest is calculated on a loan amount of $50,200 at 10% compounded semiannually 10 years. Find the ending balance of the loan.

(1141) A compound interest is calculated on a loan amount of $7,200 at 8% compounded semiannually 1 year. Find the ending balance of the loan.

(1142) A compound interest is calculated on a loan amount of $39,000 at 4% compounded semiannually 2 years. Find the ending balance of the loan.

(1143) A compound interest is calculated on a loan amount of $37,000 at 8% compounded semiannually 3 years. Find the ending balance of the loan.

(1144) A compound interest is calculated on a loan amount of $1,470 at 7% compounded semiannually 1 year. Find the ending balance of the loan.

(1145) A compound interest is calculated on a loan amount of $1,140 at 1% compounded semiannually 3 years. Find the ending balance of the loan.

(1146) A compound interest is calculated on a loan amount of $800 at 11% compounded semiannually 2 years. Find the ending balance of the loan.

(1147) A compound interest is calculated on a loan amount of $24,300 at 13% compounded semiannually 2 years. Find the ending balance of the loan.

(1148) A compound interest is calculated on a loan amount of $47,500 at 14% compounded semiannually 5 years. Find the ending balance of the loan.

(1149) A compound interest is calculated on a loan amount of $10,000 at 12% compounded semiannually 2 years. Find the ending balance of the loan.

(1150) A compound interest is calculated on a loan amount of $125 at 15% compounded semiannually 2 years. Find the ending balance of the loan.

(1151) A compound interest is calculated on a loan amount of $38,900 at 3% compounded semiannually 10 years. Find the ending balance of the loan.

(1152) A compound interest is calculated on a loan amount of $17,000 at 12% compounded semiannually 3 years. Find the ending balance of the loan.

(1153) A compound interest is calculated on a loan amount of $55 at 9% compounded semiannually 2 years. Find the ending balance of the loan.

(1154) A compound interest is calculated on a loan amount of $4,500 at 12% compounded semiannually 3 years. Find the ending balance of the loan.

(1155) A compound interest is calculated on a loan amount of $43,300 at 11% compounded semiannually 8 years. Find the ending balance of the loan.

(1156) A compound interest is calculated on a loan amount of $16,000 at 10% compounded semiannually 5 years. Find the ending balance of the loan.

(1157) A compound interest is calculated on a loan amount of $290 at 9% compounded semiannually 6 years. Find the ending balance of the loan.

(1158) A compound interest is calculated on a loan amount of $38,600 at 13% compounded semiannually 2 years. Find the ending balance of the loan.

(1159) A compound interest is calculated on a loan amount of $1,700 at 14% compounded semiannually 10 years. Find the ending balance of the loan.

(1160) A compound interest is calculated on a loan amount of $475 at 14% compounded semiannually 5 years. Find the ending balance of the loan.

PERCENTAGES
Answer key
2
3
A
B
C

Answer Key

(1) 0.264	(2) 3.125	(3) 0.646	(4) 0.47
(5) 0.55	(6) $0.\overline{3}$	(7) 0.024	(8) 0.047
(9) 0.89	(10) $0.\overline{8}$	(11) 0.63	(12) $0.\underline{31}25$
(13) 0.84	(14) 0.592	(15) 0.6	(16) 0.81
(17) 0.007	(18) 0.893	(19) 0.44	(20) 0.01
(21) 0.04	(22) 0.25	(23) 7.07	(24) $0.\overline{6}$
(25) 7.21	(26) 7.25	(27) 0.5	(28) 0.75

(29) $0.\overline{006}$	(30) 0.4	(31) 0.9	(32) $0.\overline{69}$
(33) 0.34	(34) 0.005	(35) 0.156	(36) $0.\overline{60}$
(37) $0.\overline{36}$	(38) 0.008	(39) 0.2	(40) 0.843
(41) 0.7	(42) $2.\overline{6}$	(43) 9.99	(44) $0.\overline{7}$
(45) 0.8	(46) $0.\overline{38}$	(47) $0.\overline{21}$	(48) 0.875
(49) 0.625	(50) 0.3	(51) $\frac{81}{100}$	(52) $\frac{7}{20}$
(53) $\frac{361}{500}$	(54) $\frac{207}{250}$	(55) $\frac{1}{125}$	(56) $\frac{20}{3333}$

(57) $\frac{7}{10}$ (58) $\frac{7}{8}$ (59) $\frac{43}{99}$ (60) $1\frac{1}{3}$

(61) $\frac{1}{5}$ (62) $\frac{3}{8}$ (63) $\frac{139}{303}$ (64) $\frac{341}{1000}$

(65) $\frac{1}{333}$ (66) $\frac{6}{25}$ (67) $\frac{4}{5}$ (68) $\frac{1}{8}$

(69) $\frac{1}{11}$ (70) $6\frac{9}{10}$ (71) $\frac{1}{2}$ (72) $3\frac{1}{2}$

(73) $\frac{19}{25}$ (74) $\frac{21}{100}$ (75) $\frac{1}{9}$ (76) $\frac{3}{400}$

(77) $4\frac{2}{5}$ (78) $\frac{33}{100}$ (79) $\frac{5}{9}$ (80) $\frac{3}{4}$

(81) $6\frac{1}{4}$ (82) $\frac{3}{25}$ (83) $\frac{1}{3}$ (84) $\frac{3}{10}$

(85) $\frac{2}{999}$ (86) $\frac{1}{100}$ (87) $\frac{11}{50}$ (88) $3\frac{3}{10}$

(89) $\frac{26}{33}$ (90) $\frac{2}{5}$ (91) $\frac{2}{3}$ (92) $\frac{8}{9}$

(93) $\frac{37}{100}$ (94) $\frac{3}{5}$ (95) $\frac{91}{200}$ (96) $2\frac{2}{3}$

(97) $\frac{3}{100}$ (98) $\frac{9}{10}$ (99) $\frac{1}{4}$ (100) $8\frac{7}{10}$

(101) 11% (102) 0.1% (103) 90% (104) $88.\overline{8}\%$

(105) $0.\overline{200}\%$ (106) $33.\overline{3}\%$ (107) 70% (108) 50%

(109) 70 2% (110) $666\overline{6}\%$ (111) $60\,\overline{60}\%$ (112) 212 5%

(113) $57.\overline{57}\%$	(114) 1%	(115) $76.\overline{76}\%$	(116) $36.\overline{36}\%$
(117) $18.\overline{18}\%$	(118) 87%	(119) 37.5%	(120) $42.\overline{42}\%$
(121) 20%	(122) $46.\overline{46}\%$	(123) $69.\overline{69}\%$	(124) 150%
(125) 0.4%	(126) 0.8%	(127) 73%	(128) 30%
(129) 275%	(130) 87.2%	(131) 87.5%	(132) 60%
(133) $26.\overline{426}\%$	(134) 62.5%	(135) $75.\overline{75}\%$	(136) 10%
(137) 80%	(138) 39%	(139) 40%	(140) $90.\overline{90}\%$

(141) 620%	(142) $56.\overline{56}\%$	(143) 75%	(144) $45.\overline{45}\%$
(145) 7.5%	(146) 360%	(147) 74.6%	(148) 12.5%
(149) 25%	(150) $0.\overline{300}\%$	(151) $8.\overline{08}\%$	(152) 0.5%
(153) $27.\overline{27}\%$	(154) 587 5%	(155) 725%	(156) $86.\underline{5}\%$
(157) 43%	(158) $3.\overline{03}\%$	(159) $54.\overline{54}\%$	(160) 66.6%
(161) 190%	(162) 325%	(163) 94%	(164) 49%
(165) 640%	(166) 1%	(167) 55%	(168) 250%

(169) 54%	(170) 312 %	(171) 6%	(172) 62%
(173) 97%	(174) 63%	(175) 53%	(176) 610%
(177) 99%	(178) 0%	(179) 40%	(180) 50%
(181) 1%	(182) 1%	(183) 41%	(184) 36%
(185) 11%	(186) 55%	(187) 150%	(188) 2%
(189) 75%	(190) 71%	(191) 0%	(192) 57%
(193) 1%	(194) 38 %	(195) 550 %	(196) 1%

(197) 1%	198) 1%	199) 275%	(200) 33%
(201) 0%	(202) 940%	(203) 3%	(204) 12%
(205) 95%	(206) 25%	(207) 23%	(208) 0%
(209) 70%	(210) 10%	(211) 1%	(212) 20%
(213) 77%	(214) 77%	(215) 862%	(216) 12%
(217) 60%	(218) 95%	(219) 81%	(220) 80%
(221) 96%	(222) 0.66%	(223) 0.05%	(224) 35%

(225) 90%	(226) 14%	(227) 50%	(228) 0.69%
(229) 0.58%	(230) 39%	(231) 38.3%	(232) 84%
(233) 1%	(234) 47%	(235) 35.8%	(236) 8%
(237) 2%	(238) 51%	(239) 34.3%	(240) 0.94%
(241) 0.89%	(242) 39.4%	(243) 20%	(244) 21.8%
(245) 48.7%	(246) 29%	(247) 62%	(248) 27%
(249) 6%	(250) 70%	(251) 0.37%	(252) 0.903%

(253) 10%	(254) 67%	(255) 80%	(256) 45.3%
(257) 7%	(258) 99.3%	(259) 0.88%	(260) 93.2%
(261) 60%	(262) 9%	(263) 52%	(264) 40%
(265) 38%	(266) 4%	(267) 0.23%	(268) 75%
(269) 16.7%	(270) 34.7%	(271) 5%	(272) 88%
(273) 0.305%	(274) 51.8%	(275) 77%	(276) 21%
(277) 81%	(278) 18%	(279) 63%	(280) 2.1%

(281) 186%	(282) 42%	(283) 94%	(284) 32%
(285) 3%	(286) 98%	(287) 66%	(288) 758%
(289) 16%	(290) 106%	(291) 9%	(292) 259%
(293) 221%	(294) 20%	(295) 80%	(296) 69%
(297) 85%	(298) 64%	(299) 36%	(300) 88%
(301) 77%	(302) 779%	(303) 333%	(304) 53%
(305) 50%	(306) 388%	(307) 488%	(308) 163%

(309) 40%	(310) 209%	(311) 90%	(312) 324%
(313) 17%	(314) 11%	(315) 962%	(316) 59%
(317) 52%	(318) 1%	(319) 690%	(320) 27%
(321) 5%	(322) 477%	(323) 2%	(324) 22%
(325) 45%	(326) 495%	(327) 10%	(328) 91%
(329) 6%	(330) 19%	(331) 244%	(332) 161%
(333) 15%	(334) 44%	(335) 4%	(336) 87%

(337) 41%	(338) 510%	(339) 89%	(340) 509%
(341) 81%	(342) 94.9%	(343) 198.4%	(344) 53.2%
(345) 48.7%	(346) 38.9%	(347) 251.1%	(348) 75%
(349) 7.7%	(350) 123%	(351) 179.2%	(352) 175%
(353) 71.9%	(354) 624%	(355) 40.7%	(356) 10.8%
(357) 24.3%	(358) 89.4%	(359) 85%	(360) 200%
(361) 90.2%	(362) 85.5%	(363) 106.5%	(364) 169.4%

(365) 92.6%	(366) 26.7%	(367) 37.8%	(368) 75.9%
(369) 69.9%	(370) 0.89%	(371) 83%	(372) 9.5%
(373) 45.9%	(374) 87%	(375) 72.7%	(376) 29.6%
(377) 73.4%	(378) 1.54%	(379) 40.8%	(380) 55%
(381) 15%	(382) 49.1%	(383) 97.1%	(384) 100%
(385) 39.6%	(386) 3.2%	(387) 18.4%	(388) 13.6%
(389) 38.3%	(390) 31.1%	(391) 71.6%	(392) 1.53%

(393) 29.9%	(394) 23.8%	(395) 85.2%	(396) 40%
(397) 390.5%	(398) 13.2%	(399) 235.3%	(400) 29.2%
(401) 106%	(402) 53.5%	(403) 8.5%	(404) 9.8%
(405) 85.8%	(406) 3.5%	(407) 158.2%	(408) 13.3%
(409) 56.3%	(410) 64.2%	(411) 1.35%	(412) 11.3%
(413) 168.3%	(414) 34%	(415) 573.9%	(416) 20.6%
(417) 49.2%	(418) 22%	(419) 20.6%	(420) 24.8%

(449) 110 ft	(450) 39.6 tons	(451) 48.3 inches	(452) $126.42
(453) 134.4 ft	(454) 2.7 tons	(455) 153.2 miles	(456) 100.5 ft
(457) 4.8 grams	(458) $54.76	(459) 11.6 tons	(460) 392.6 tons
(461) 9.4 minutes	(462)147.3 minutes	(463) 100.6 grams	(464) 180.2 tons
(465) 29.8 inches	(466) 75.7 grams	(467) 26.5 tons	(468) 336 tons
(469) 329 m	(470) 3.9 minutes	(471) $158.40	(472) $7.68
(473) 4.3 m	(474) 34.3 grams	(475) 121.5 minutes	(476) 20.5 miles

(421) 417.6 ft	(422) 76.4 inches	(423) 56.1 minutes	(424) 8.8 hours
(425) 216 hours	(426) 82.9 tons	(427) 17.5 m	(428) 40.4 minutes
(429) 45.2 ft	(430) 0.8 hours	(431) 26.6 tons	(432) 45.7 grams
(433) 19.8 tons	(434) 28.1 grams	(435) 107.9 inches	(436) 53.5 grams
(437) 5.9 tons	(438) $7.04	(439) 47.4 grams	(440) 132.8 inches
(441) 102.9 m	(442) 2.2 hours	(443) 160.7 m	(444) $0.36
(445) 334.8 minutes	(446) 22 grams	(447) $86.40	(448) 2.3 m

(477) 19.3 inches	(478) 0.87 ft	(479) 15.7 minutes	(480) 30.5 m
(481) 77.4 minutes	(482) 52.9 m	(483) 1.44 inches	(484) 77.8 minutes
(485) 10.9 inches	(486) 3.2 grams	(487) $98.40	(488) 23.4 tons
(489) $5.55	(490) 421.8 m	(491) 125.5 miles	(492) 15.8 inches
(493) 278.4 ft	(494) 30.8 m	(495) 292.6 minutes	(496) $92.40
(497) 77.1 inches	(498) 12.5 tons	(499) 26.2 miles	(500) 84.6 miles
(501) 140	(502) 283	(503) 7.8	(504) 231

(505) 62.4	(506) 110.3	(507) 41.1	(508) 105.2
(509) 141.1	(510) 172.9	(511) 236	(512) 485.7
(513) 194.9	(514) 940	(515) 150.7	(516) 191.1
(517) 39.4	(518) 9400	(519) 14.8	(520) 7.3
(521) 432.6	(522) 71.4	(523) 64.4	(524) 151.9
(525) 206.8	(526) 26.3	(527) 133.3	(528) 55.7
(529) 257.8	(530) 62.7	(531) 97.2	(532) 17.7

(533) 77.3	(534) 75.4	(535) 85	(536) 33.6
(537) 29	(538) 1866.7	(539) 2940	(540) 1633.3
(541) 8.4	(542) 707.1	(543) 99.1	(544) 108.8
(545) 413.9	(546) 1153.8	(547) 81.3	(548) 141.5
(549) 300	(550) 91.2	(551) 68.7	(552) 11.3
(553) 560	(554) 77.7	(555) 220	(556) 160.6
(557) 278.9	(558) 105.6	(559) 10.8	(560) 156.8

(561) 571.7	(562) 24.1	(563) 189.6	(564) 36
(565) 49.5	(566) 56.6	(567) 87.7	(568) 121.3
(569) 271.5	(570) 52.9	(571) 46	(572) 4.2
(573) 12.3	(574) 300	(575) 2650	(576) 126.9
(577) 8.2	(578) 88	(579) 28	(580) 30.3

(581) 193.6% increase

(582) 6.7% decrease

(583) 118.1% increase

(584) 295.5% increase

(585) 35.3% decrease

(586) 312.5% increase

(587) 92.8% decrease

(588) 32.8% increase

(589) 322.2% increase

(590) 84.9% decrease

(591) 87.2% decrease

(592) 129.2% increase

(593) 175.9% increase

(594) 27% decrease

(595) 50.3% decrease

(596) 34.2% decrease

(597) 4.3% increase

(598) 89.1% decrease

(599) 15.7% increase

(600) 56.1% increase

(601) 21.6% decrease

(602) 70.1% decrease

(603) 56.7% increase

(604) 238.5% increase

(605) 28.2% increase

(606) 89.8% increase

(607) 273.6% increase

(608) 3.5% increase

(609) 134% increase

(610) 18.4% decrease

(611) 47.6% decrease

(612) 6.1% decrease

(613) 4.5% increase

(614) 92.2% decrease

(615) 163.5% increase

(616) 64.1% decrease

(617) 25% decrease

(618) 49.6% increase

(619) 39.4% increase

(620) 8.5% decrease

(621) 4.7% decrease

(622) 277.5% increase

(623) 70.7% decrease

(624) 211.1% increase

(625) 68.8% decrease

(626) 7.2% increase

(627) 73.7% increase

(628) 58.7% decrease

(629) 40.8% increase

(630) 58.4% decrease

(631) 51.3% decrease

(632) 88.6% decrease

(633) 24.2% decrease

(634) 24.9% decrease

(635) 25.1% decrease

(636) 39.1% increase

(637) 80% decrease

(638) 98.2% decrease

(639) 5.4% increase

(640) 97% decrease

(641) 79.7% decrease

(642) 51.7% decrease

(643) 206.1% increase

(644) 18% decrease

(645) 20.6% decrease

(646) 3.5% increase

(647) 95.8% decrease

(648) 86.6% decrease

(649) 36.8% decrease

(650) 89.6% increase

(651) 8.2% decrease

(652) 60.9% increase

(653) 58.7% increase

(654) 39.8% decrease

(655) 44.2% increase

(656) 198.1% increase

(657) 95.3% decrease

(658) 66.3% decrease

(659) 93.7% increase

(660) 288% increase

(661) 35.7% increase

(662) 56% decrease

(663) 177.8% increase

(664) 40.3% decrease

(665) 2.3% increase

(666) 18% increase

(667) 58.7% increase

(668) 212.5% increase

(669) 56.8% decrease

(670) 300% increase

(671) 76.9% decrease

(672) 1.4% increase

(673) 15.4% increase

(674) 8.7% decrease

(675) 65.3% decrease

(676) 90.6% decrease

(677) 150% increase

(678) 28.6% decrease

(679) 48.2% increase

(680) 16.7% decrease

(681) 95.7% decrease

(682) 15.4% decrease

(683) 30.8% increase

(684) 77.8% decrease

(685) 58.6% decrease

(686) 269.9% increase

(687) 43.5% decrease

(688) 15.6% increase

(689) 24.2% increase

(690) 63.9% increase

(691) 133.3% increase

(692) 23.8% decrease

(693) 76% increase

(694) 21.5% decrease

(695) 21.3% increase

(696) 68% decrease

(697) 23.9% increase

(698) 71.8% decrease

(699) 32.9% decrease

(700) 159.4% increase

(701) 150% increase

(702) 63.6% decrease

(703) 25.9% increase

(704) 22.6% increase

(705) 31.5% decrease

(706) 17.9% decrease

(707) 41.2% increase

(708) 13.7% increase

(709) 14% increase

(710) 16.1% decrease

(711) 41.2% decrease

(712) 63.2% decrease

(713) 58.5% increase

(714) 292.9% increase

(715) 9.1% increase

(716) 377.8% increase

(717) 40.9% increase

(718) 10.3% decrease

(719) 63.8% decrease

(720) 79.8% decrease

(721) 314.3% increase

(722) 58.2% increase

(723) 20.5% decrease

(724) 58% decrease

(725) 91.2% increase

(726) 105.6% increase

(727) 238.9% increase

(728) 50.3% decrease

(729) 85.1% decrease

(730) 84.5% decrease

(731) 87.3% decrease

(732) 96.9% decrease

(733) 31% decrease

(734) 20.3% increase

(735) 1.1% increase

(736) 44.3% decrease

(737) 258.3% increase

(738) 39.6% decrease

(739) 90.9% decrease

(740) 237% increase

(741) $6.39

(742) $55.99

(743) $45.47

(744) $10.00

(745) $1.50

(746) $14.97

(747) $151.19

(748) $159.99

(749) $10.32

(750) $224.98

(751) $175.00

(752) $0.45

(753) $52.15

(754) $54.36

(755) $500.00

(756) $97.99

(757) $3,796.00

(758) $70.00

(759) $67.89

(760) $89.99

(761) $3.20

(762) $91.47

(763) $151.99

(764) $36,045.00

(765) $29.10

(766) $27.50

(767) $100.50

(768) $139.50

(769) $30.50

(770) $16.14

(771) $17.99

(772) $51.96

(773) $12,500.00

(774) $119.70

(775) $52.49

(776) $33.91

(777) $2.07

(778) $7.96

(779) $14.93

(780) $40.95

(781) $20.00

(782) $3.90

(783) $439.60

(784) $172.46

(785) $62.96

(786) $324.07

(787) $10.50

(788) $3.99

(789) $40.57

(790) $174.65

(791) $104.49

(792) $74.75

(793) $2,100.00

(794) $12.38

(795) $34.97

(796) $1.02

(797) $266.00

(798) $80.00

(799) $120.00

(800) $4.28

(801) $74.99

(802) $6,000.00

(803) $26.78

(804) $97.89

(805) $87.96

(806) $15.99

(807) $47.48

(808) $1.79

(809) $18.90

(810) $1.18

(811) $55.65

(812) $311.96

(813) $288.59

(814) $27.96

(815) $50.00

(816) $22.04

(817) $38.00

(818) $462.41

(819) $447.41

(820) $40.79

(821) $54.45

(822) $1,560.00

(823) $25,500.00

(824) $15,840.00

(825) $54.00

(826) $72,000.00

(827) $55,100.00

(828) $49.45

(829) $49,500.00

(830) $110.48

(831) $41.43

(832) $23,100.00

(833) $1,714.13

(834) $50.39

(835) $164.45

(836) $81.75

(837) $389.99	(838) $20.93
(839) $61.12	(840) $53.55
(841) $3,240.00	(842) $52.43
(843) $34.26	(844) $2,700.00
(845) $19.58	(846) $231.23
(847) $179.99	(848) $2.99
(849) $173.28	(850) $62.41

(851)	$309.94	(852)	$119.99
(853)	$8.57	(854)	$142.95
(855)	$49,500.00	(856)	$6.93
(857)	$3,363.00	(858)	$8.40
(859)	$132.13	(860)	$164.99
(861)	$41.40	(862)	$14.32
(863)	$39.83	(864)	$2.75

(865) $85.41

(866) $33.25

(867) $3.00

(868) $157.49

(869) $13.15

(870) $17.10

(871) $610.50

(872) $7.74

(873) $399.92

(874) $364.00

(875) $25.94

(876) $22.80

(877) $255.00

(878) $119.40

(879) $65.18

(880) $29.03

(881) $95.19

(882) $59.99

(883) $16.53

(884) $216.78

(885) $3,000.00

(886) $25.38

(887) $7.72

(888) $12,730.00

(889) $12,760.00

(890) $65,800.00

(891) $139.00

(892) $23.40

(893) $17.85

(894) $56.56

(895) $45.40

(896) $9.69

(897) $17.33

(898) $77.20

(899) $29.80

(900) $20.80

(901) $2.55

(902) $93.08

(903) $159.00

(904) $176.80

(905) $155.48

(906) $20.15

(907) $11.54

(908) $8.76

(909) $428.40

(910) $2.53

(911) $307.40

(912) $36.74

(913) $253.87

(914) $14.79

(915) $63.07

(916) $167.95

(917) $5,355.00

(918) $395.15

(919) $79.04

(920) $36,570.00

(921) $20.39

(922) $10.07

(923) $15.23

(924) $4.15

(925) $4.24

(926) $155.99

(927) $31.15

(928) $52.47

(929) $11.32

(930) $2,862.00

(931) $49.35

(932) $423.25

(933) $1.99

(934) $87.04

(935) $91.75	(936) $105.95
(937) $51.48	(938) $19.75
(939) $51.50	(940) $314.48
(941) $1,470.00	(942) $26.24
(943) $404.00	(944) $16,800.00
(945) $25.76	(946) $644.80
(947) $42.35	(948) $207.95

(949) $306.00

(950) $33,330.00

(951) $46.30

(952) $2.58

(953) $101.99

(954) $31.49

(955) $9.26

(956) $259.99

(957) $113.30

(958) $1.01

(959) $836.40

(960) $24,205.00

(961) $494.40

(962) $60.69

(963) $66.94	(964) $32.24
(965) $20.48	(966) $25,404.00
(967) $2,004.00	(968) $277.20
(969) $336.00	(970) $693.00
(971) $856.80	(972) $15,600.00
(973) $1,529.00	(974) $8,064.00
(975) $36,270.00	(976) $560.00

(977)	\$13,284.00	(978)	\$49,184.00
(979)	\$34,320.00	(980)	\$33,600.00
(981)	\$279.40	(982)	\$3,330.00
(983)	\$4,240.00	(984)	\$493.00
(985)	\$660.00	(986)	\$89.60
(987)	\$1,134.20	(988)	\$50,700.00
(989)	\$52,200.00	(990)	\$520.20

(991) $168.00

(992) $2,420.00

(993) $832.00

(994) $1,636.80

(995) $195.75

(996) $35,687.00

(997) $1,260.00

(998) $79,464.00

(999) $403.30

(1000) $14,560.00

(1001) $510.30

(1002) $70,400.00

(1003) $13,920.00

(1004) $278.00

(1005) $31,320.00

(1006) $1,776.00

(1007) $98,968.00

(1008) $43.20

(1009) $559.00

(1010) $5,330.00

(1011) $70,520.00

(1012) $436.80

(1013) $1,009.20

(1014) $1,092.50

(1015) $24,990.00

(1016) $34,880.00

(1017) $48,360.00

(1018) $35,340.00

(1019) $50,386.00 (1020) $65,780.00

(1021) $22,836.00 (1022) $8,800.00

(1023) $33,000.00 (1024) $21,901.00

(1025) $9,450.00 (1026) $11,088.90

(1027) $480.52 (1028) $2,524.22

(1029) $54,138.10 (1030) $72,647.60

(1031) $67,642.29 (1032) $417.47

(1033) $1,516.55	(1034) $258.74
(1035) $11,171.99	(1036) $67,189.32
(1037) $62,929.63	(1038) $2,449.04
(1039) $457.95	(1040) $883.30
(1041) $5,202.00	(1042) $136,215.86
(1043) $1,648.48	(1044) $258.55
(1045) $1,771.56	(1046) $861.18

(1047) $60,786.76

(1048) $80,878.90

(1049) $1,321.31

(1050) $2,361.12

(1051) $57,689.94

(1052) $65,259.30

(1053) $426.80

(1054) $116,455.54

(1055) $1,359.68

(1056) $15,322.80

(1057) $308.91

(1058) $26.52

(1059) $14,204.18

(1060) $34,476.30

(1061) $632.23

(1062) $7,571.20

(1063) $634.64

(1064) $1,575.71

(1065) $21,180.65

(1066) $62,781.07

(1067) $84,052.93

(1068) $851.00

(1069) $57,378.53

(1070) $29,503.49

(1071) $111,918.30

(1072) $1,651.46

(1073) $1,819.58

(1074) $151.63

(1075) $790.08

(1076) $668.36

(1077) $3,138.44

(1078) $63,041.00

(1079) $69,085.60

(1080) $2,697.46

(1081) $822.03

(1082) $160.17

(1083) $54,837.12

(1084) $15,791.18

(1085) $101,442.36

(1086) $37,208.13

(1087) $203.09

(1088) $2,021.56

(1089)	$67,434.61	(1090)	$44,577.27
(1091)	$1,183.55	(1092)	$28,170.63
(1093)	$1,109.28	(1094)	$585.57
(1095)	$49,929.94	(1096)	$72,210.91
(1097)	$30,833.48	(1098)	$4,502.63
(1099)	$102.01	(1100)	$50,715.00
(1101)	$11,670.72	(1102)	$15,661.74

(1103) $94,423.27 (1104) $304.16

(1105) $425.43 (1106) $67,486.12

(1107) $1,659.80 (1108) $20,197.52

(1109) $6,491.38 (1110) $115,327.41

(1111) $48,177.75 (1112) $55,174.76

(1113) $692.76 (1114) $970.98

(1115) $1,040.71 (1116) $1,898.01

(1117) $5,157.07

(1118) $223.77

(1119) $31,668.73

(1120) $23,149.52

(1121) $1,290.47

(1122) $57,323.07

(1123) $43,691.89

(1124) $5,852.65

(1125) $1,149.27

(1126) $33,153.25

(1127) $3,761.87

(1128) $382.70

(1129) $11,041.45

(1130) $235.53

(1131) $27,595.45

(1132) $2,217.50

(1133) $32,033.01

(1134) $4,642.16

(1135) $136.35

(1136) $52,572.32

(1137) $1,887.52

(1138) $1,144.66

(1139) $191.56

(1140) $133,195.54

(1141) $7,787.52

(1142) $42,214.85

(1143) $46,816.80

(1144) $1,574.70

(1145) $1,174.63

(1146) $991.06

(1147) $31,261.13

(1148) $93,439.69

(1149) $12,624.77

(1150) $166.93

(1151) $52,392.66

(1152) $24,114.82

(1153) $65.59

(1154) $6,383.34

(1155) $101,982.87

(1156) $26,062.31

(1157) $491.81

(1158) $49,657.60

(1159) $6,578.46

(1160) $934.40

Made in United States
Orlando, FL
05 November 2026

90229681R00157